TH
638
M38

W9-AEH-174

HEAT PUMPS

An Efficient Heating
& Cooling Alternative

Dermot McGuigan

with Amanda McGuigan

GARDEN WAY  PUBLISHING
CHARLOTTE, VERMONT 05445

KVCC KALAMAZOO VALLEY
COMMUNITY COLLEGE
LIBRARY

52307

Illustrations by Bob Vogel

Copyright © 1981 by Garden Way, Inc.

All rights reserved. No part of this book may be reproduced without permission in writing from the publisher, except by a reviewer who may quote brief passages or reproduce illustrations in a review with appropriate credit; nor may any part of this book be reproduced, stored in a retrieval system, or transmitted in any form or by any means — electronic, photocopying, recording, or other — without permission in writing from the publisher.

Library of Congress Cataloging in Publication Data

McGuigan, Dermot.
   Heat pumps.

   Includes index.
   1. Heat pumps. I. McGuigan, Amanda. II. Title.
TH7638.M38   697   81–7185
ISBN 0-88266-255-4   AACR2

ii

Contents

Special Thanks

I am indebted to my wife, Amanda, and to Neville Rowe for helping me with this book.

Of the many I met in my travels, I want to thank in particular the following: Julian Keable, Winton Goodchild (W. R. Heat Pumps Ltd.), John Brady (National Science Council), John Brown, Gordon Fabian (Vanguard Energy Systems), Gale Carson, and John Sumner.

THE HEAT PUMP AND THE SATYR

Aesop tells the tale of a traveler who sought refuge with a satyr on a bitter winter night. On entering the satyr's lodging, the traveler blew on his fingers. The satyr asked the traveler why he did this. "To warm them up," the traveler explained.

Later, on being served a piping hot bowl of porridge, the traveler blew on it. Puzzled, the satyr asked why he now blew on the porridge. "To cool it," was the reply. Thereupon the satyr thrust the traveler out-of-doors, for he would have nothing to do with a man who could blow hot and cold with the same breath.

***　　　　***　　　　***

Heat pumps blow both hot and cold, but better than that they give out far more in heat than they take in energy. In this they are generous but definitely not intended for satyrs.

Introduction

Heat pumps save energy by tapping into an energy source that is universally available. The source is around the home, in the air, the sun, the earth, and the water. The energy is free; the only cost is the cost of collection.

We go about our daily lives in an abundance of low temperature heat. Even when the air is below 32° F., it contains vast reserves of heat that can be used by a heat pump. The real excitement and potential of heat pumps are that for each unit of primary energy — gas, oil, coal, nuclear, hydroelectricity, wind power, fuel alcohol, and even solar power — the heat pump *multiplies* the primary energy by two, three, or four times, and even in some cases by as much as ten times. Heat pumps can, by taking one kilowatt of electricity, transform that single unit into anything from 1½ to 10 times as much in heat equivalent. Can you see the importance and potential for utilizing these appliances in an age when the energy shortage has become a reality? A consideration of heat pumps makes us realize we have no shortage of energy; all we have is an addiction to the use of certain forms of energy.

Pumping heat

Heat pumps, if used to their fullest potential, can vastly reduce the consumption of what is left of our fossil fuel reserves. *Heat pumps transfer heat and in so doing, increase the temperature of that heat by "compressing" it to a higher temperature suitable for use in home and industry.* A heat pump can take a few degrees of heat from the air at a winter temperature of 30° F. and raise that temperature to 100° F. to be distributed for home heating.

Looked at another way, *heat pumps multiply heat.* In fact, the heat pump was first called a "heat multiplier" by the man who invented it, Lord Kelvin. In the heat pump, it is the compressor that raises or "multiplies" the low temperature heat to a temperature that is useful for heating.

The opportunities for the use of heat pumps in the next decade are great. Since 1973, the rise in the price of fossil fuels and even renewable fuels such as wood, has opened the way for the mass application of heat pump technology to existing and new houses. In commercial and agricultural plants, there is almost unlimited potential for new adaptations of the heat pump and for increased sales of current models.

And what does this mean to you, the homeowner? Whether a heat pump is for you will depend on many factors — your present heating system, your air-conditioning needs, and the like. The Oak Ridge National Laboratory gave an indication of what to expect in a statement prepared for the Department of Energy: "While actual savings depend upon such factors as climate and the price of . . . electricity, heat pumps offer an average of 20 percent savings over conventional cooling systems and central electric resistance heating . . . . In some regions of the country a heat pump can reduce electric bills by 35 to 45 percent."

Energy supply

We import about 50 percent of our oil. However 80 percent of all the energy we use — coal, gas, hydro, nuclear, and oil — is produced within this country. We have all the energy we need if what we have is used wisely. Heat pumps alone have the ability to make up most of the 20 percent difference simply by making use of the energy that is all around us and within our control.

America can and needs to achieve energy independence. By using what fossil and renewable energy sources we have today in conjunction with heat pumps, we can both extend the life of our reserves and decrease our debilitating dependence on the vagaries of the oil-producing nations.

By using heat pumps we can save money, save energy, and therefore contribute to a stronger, safer America. The purpose of this book is to show how you can use heat pumps to your own benefit.

PART I

HEAT
PUMPS
AND
HEATING

"We live in a blanket of low temperature heat energy — in the air, the earth, the water."

How They Began

Heat pumps and refrigerators are of the same family, follow the same principles, and are born of the same history.

In 1850, in Apalachicola, Florida, John Gorrie invented the first air-conditioning unit. In the same year Lord Kelvin, an Irishman, expounded the theory of his heat multiplier to an audience in London at the Royal Society (of which he later became president). Kelvin's "heat multiplier" is what we now call the heat pump.

It had long been known that the compression of a gas, such as air, caused its temperature to rise. Both Kelvin and Gorrie used this principle to change the temperature of working fluids.

Gorrie's preoccupation was with ice-making because ice was costing him up to $1.25 a pound to be shipped down from Massachusetts. Gorrie needed the ice to keep his malaria patients cool, which was how he hoped to cure them of their ills. Determined to avoid that cost, Gorrie fashioned his own ice maker. He built a steam engine that drove a piston to compress air, and then, quickly allowing the compressed air to expand, he succeeded in reducing the temperature of the air sufficiently to use it to freeze water. One can feel the same cooling effect when filling a gas cigarette lighter. The liquified gas rapidly expands as it goes from the canister to the lighter.

Gorrie was elated with his first batch of man-made ice. He

patented his invention, which he considered to be a success. The *New York Times* called him a crank, ridiculed his machine, and said that only the Almighty could make ice. The luckless Gorrie could not find an investor to back his product. It was not until 1890, long after Gorrie's death, that the ice-making machine went into production as a result of a shortage of natural ice.

Lord Kelvin, widely known for the Kelvin temperature scale, displayed wisdom in knowing that the market for his heat multiplier would not arrive until the world's fossil fuels were almost gone. He reasoned that as stocks of high-grade fuel diminished, the cost would go up. Only then would we begin to use what was left to upgrade the low-temperature heat found abundantly in the sun, air, earth, and water. His prediction is proving to be correct. By conserving our stock of fossil fuels and using them to drive efficient heat pumps, we can at least double the life expectancy of our reserves.

The first automatic refrigerator was called the Kelvinator and came on the market in 1918. The first heat pump was made by T.G.N. Haldane in Scotland in 1926. It was powered by a small hydroelectric plant and was used to extract about five kilowatts from tap water and air. Haldane got a return of between two to three times the energy his heat pump used — excellent, especially considering it was the first one made.

Changing market

The more recent history of heat pumps has been a story of a rapidly expanding market, a near-collapse of that market, and then, with the sudden spurt in fuel costs, renewed interest.

Heat pumps for homes were commercially produced in ever-increasing numbers in the early 1950s, and until the mid-60s. Many of these models were patterned after air-conditioning units, and were rushed into production without basic engineering tests. They began to develop problems. Compressors were unable to carry the larger load demands in the heating mode. Too, the coils iced up, reducing their efficiency. And as owners tried to get repairs, they found the industry lacked trained service personnel.

Sales dropped

The result was predictable — the popularity of heat pumps plummeted, and sales fell off, driving manufacturers and dealers out of the business.

The turnaround came with the oil embargo of 1973. As oil prices soared, major manufacturing firms, intent on avoiding the problems of the past, put money into both the design of heat pumps and the training of service personnel.

The wave of popularity for these second-generation heat pumps began with air-to-air units sold in warm areas of the country, particularly Florida and the Southwest, where the demand was for energy-efficient equipment that would both heat and cool a home. As the efficiency of heat pumps has increased, their use has spread into the colder regions. Today, more than three million heat pumps are in use in this country, heating and cooling homes, motels, factories, schools, and offices.

Today manufacturers such as York subject their heat pumps to long and rigorous testing to avoid the problems that hit the industry earlier.

Understanding the
Heat Pump

Think of the heat pump as it relates to two other household appliances, the refrigerator and the air conditioner. They're all much the same. Both of these take heat *from* one place where it is not wanted (inside the refrigerator or in a warm room) and move it to another place (into the kitchen or outside the house). The only difference between these and the heat pump is that the latter moves heat *to* where it is needed. And even this difference is eliminated when the action of the heat pump is reversed (most heat pumps can be reversed) so that it acts as an air conditioner, moving heat out of the house, instead of into it, during times when cooling is required.

The magic of the heat pump becomes apparent when we see how it operates. Put a given amount of energy into a heat pump and you get back in heat that amount of energy — and more. If you have conventional electric resistance heating, you know that for $1 worth of electricity you get back $1 in heat. But put $1 worth of electricity into the heat pump and you may get 1.5 to 10 times as much heat. Conversion from resistance heating to a heat pump can mean a cut of 35 to 45 percent in most utility bills.

How does this work? To understand the heat pump, we must think a bit differently about "heat." Let's think about it the way Lord Kelvin did, that there is some "heat" in any matter until it

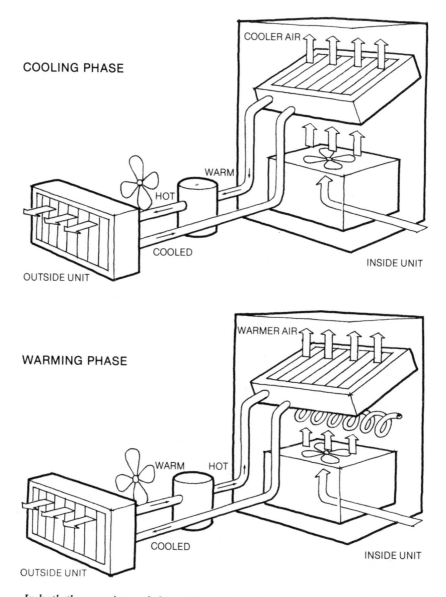

COOLING PHASE

COOLER AIR

WARM

HOT

COOLED

OUTSIDE UNIT

INSIDE UNIT

WARMING PHASE

WARMER AIR

WARM   HOT

COOLED

OUTSIDE UNIT

INSIDE UNIT

In both the warming and the cooling phases, the refrigerant picks up heat (outside when heating, inside when cooling), is compressed, gives up heat when air is blown across it, then completes the cycle as it returns to the starting point. Note flow direction in both phases.

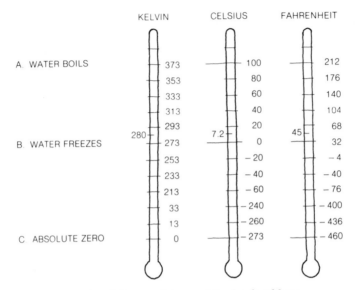

Three thermometers give different pictures of the levels of heat.

reaches the low reading of minus 460° F. He set up his own scale of temperatures. It looks like this:

KELVIN (DEGREES)		CENTIGRADE (DEGREES)		FAHRENHEIT (DEGREES)		
0° K.	=	−273° C.	=	−460° F.	=	absolute zero
273° K.	=	0° C.	=	32° F.	=	freezing point of water
373° K.	=	100° C.	=	212° F.	=	boiling point of water

Looking at these figures as Lord Kelvin did, you see that at levels of temperature we think of as cold, such as 32° F., there is still heat in the atmosphere.

What's ahead

We'll see dramatic progress in the next decade in heat pump development. Currently there is research going on in many areas, among them:

1. Solar installations, to combine the good points of solar systems and heat pumps

2. Utilizing "waste" heat in commercial buildings
3. Storage of heat
4. Improvement of components of the heat pump
5. Devising ways to use a broad array of sources of free, low-temperature heat.

Research is emphasizing ways to make the input of energy, such as electricity, as small as possible. Exciting, new methods are being developed to store heat, so that the up-and-down heat output of a unit such as a solar collector can be tempered by storing excess heat to be used later during the down-output phase of the unit.

In our efforts to produce inexpensive energy, we are moving away from a single heat source to the use of multiple sources. In some northern climates, for example, the heat pump is most efficiently used in conjunction with an alternate fuel. Similarly, in the development of heat pumps, we may be moving from using only an air-to-air heat pump to using a variety of pumps, such as water-to-air, and selecting the system best suited for an individual situation.

Blanket of heat

This makes us realize that we live in a blanket of low temperature heat energy — in the air, the earth, the water. Much of the time, such as in the summer, that blanket is warm enough so that we don't need additional heat. And even in the winter, the percentage of heat that we need added is very low. For example, even when it is freezing cold at 32° F., the air contains 91 percent of the energy required to keep us warm at 80° F. All we need to do is to find that extra 9 percent. Here's how that is figured:

$$\frac{\text{Heat at 32° F.}}{\text{Heat at 80° F.}} = \frac{273° \text{ K.}}{300° \text{ K.}} = 91\%$$

With the usual heating systems, we burn fossil fuels at very high temperatures to get a temperature increase of just a few degrees in the air around us. This involves a great deal of waste since the average domestic burner, gas- or oil-fired, operates at an efficiency of about 50 to 60 percent. And if you heat with electricity, the heating within your home is very efficient, but the generator-distribution system usually has an efficiency of only about 35 percent.

The alternative, the heat pump, takes heat from the blanket around us, upgrades it a few degrees, then releases it in our homes.

Other principles

There are several other principles and terms you should understand before we probe into that mysterious box that is the heat pump.

These are:

1. Heat flows from a warm surface to a cold surface.
2. A refrigerant is a liquid that boils at a very low temperature, far below the 212° F. boiling point of water.
3. When a liquid such as a refrigerant boils and becomes a vapor, it absorbs heat.
4. When it changes back from a vapor to a liquid, it gives up that heat.
5. Both pressure and temperature influence whether a refrigerant is a gas or liquid.

You'll see all of this information used in the mechanics of the heat pump.

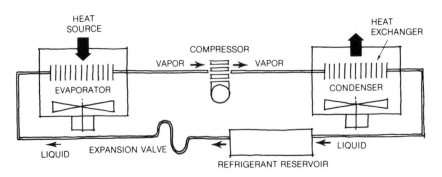

When refrigerant absorbs heat, it changes from a liquid to a vapor.

A Look at a Heat Pump

The first heat pump you see will probably be a conventional model, which takes heat from air outside the home and moves it into the house, where it is distributed through a conventional system of ducts, much like any hot-air system.

You'll see a metal box outside the home, a similar box inside, and pipes and wires connecting the two.

The outside box contains a heat exchange coil, a compressor run

Here's the entire package of the Weathermaster III heat pump model 38TQ, including the outside unit, at left, and inside units.

This single package, an Amana unit, contains an entire heat pump.

The interior of a similar unit, a Singer heat pump, showing the compressor, lower left, the controls, air inlet and fan.

by electricity, and a reversing valve. The inside box, which looks much the same, contains the heat exchange coil, the fan, and controls. A refrigerant (Freon 12 is a common one, boiling at below 0° F.) moves into the outside box in a cool state. Air is blown across it so that it can pick up heat, and then it is compressed to raise the level of that heat, flows into the house, and gives off that heat.

Steps of operation

Let's take that step by step, in just a bit more detail.

1. The refrigerant flows to the outdoor box. In the heating mode the refrigerant absorbs heat from the air around the outdoor coil. This causes the refrigerant to evaporate into a low-pressure gas.
2. The gas is pumped into the compressor, where it is compressed into a hot, high-pressure vapor, about 100° F.
3. It travels through a pipe into the box in the house, goes into the indoor coil — a condenser — and gives off its heat as air is fanned across the coil.

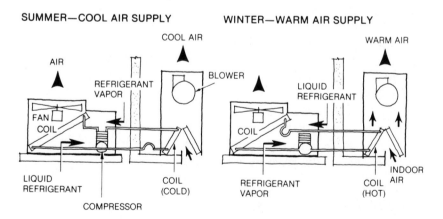

In summer mode, this heat pump pushes warmed air from house through compressor, so that it is heated at a higher level before reaching coil outside. In winter mode, flow direction of refrigerant is reversed, and the refrigerant picks up heat outside, then passes through compressor and goes to inside coil.

WINTER
1 COLD INDOOR AIR RETURN
2 COOLED REFRIGERANT OUT
3 REFRIGERANT HEATED
4 COLD OUTDOOR AIR ENTERING
5 COLDER OUTDOOR AIR DISCHARGE
6 HEATED REFRIGERANT IN
7 HEATED INDOOR AIR TO HOUSE

SUMMER
1 HOT INDOOR AIR RETURNED
2 HEATED REFRIGERANT OUT
3 REFRIGERANT COOLED
4 WARM OUTDOOR AIR ENTERING
5 HOT OUTDOOR AIR DISCHARGE
6 COOLED REFRIGERANT IN
7 COOLED INDOOR AIR TO HOUSE

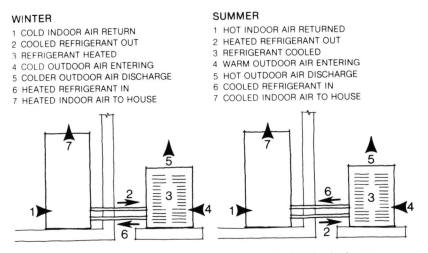

Details of heat pump cycle, from house, left, to unit outside the house.

4. Loss of the heat turns the vapor back into a liquid, and the pressure is then reduced through an expansion valve. The refrigerant, cold and not under pressure, returns to the outdoor coil, and the cycle continues.

Runs in reverse

Put this entire system into reverse, through a four-way valve, and the cycle is reversed, with the heat pump removing heat from the house during hot weather.

The heat pump cycle also reverses for one more function. When the outside temperature drops much below 45° F. and the humidity is high, frost will form on the outside coil. Most heat pumps reverse when that frost buildup reaches excessive amounts. The cycle to remove the frost lasts for up to ten minutes.

Advantages and disadvantages

As with all heating systems, the heat pump has its advantages and disadvantages, compared with other heating systems. Let's run down through some of them.

Advantages
1. The heat pump is cheaper to operate, given a situation where its cost-effectiveness has been studied in advance. In general, it is most efficient, and provides the best service for its owner, in situations where the number of degrees the temperature must be raised is least.
2. One heat pump supplies both heat and cooling, and only one system of distribution is needed for both.
3. A heat pump is clean and safe. There's no wood to chop, store, and lug into the home; no ashes to remove; no coal or oil to store; no danger of fumes and gases; no fire to spread from the heating unit.
4. A heat pump can be made to use any constant supply of low-grade heat.

Disadvantages
1. The initial cost may be from 25 to 50 percent greater than an oil or gas furnace, but the difference in cost will be less if the price of an air-conditioning system is added to that of the furnace.
2. Some units are noisy. Careful selection and placement of a unit, and simple methods of deadening sound can combine to avoid this problem.
3. The heat pump acts exactly in reverse from what would be ideal. As the weather cools and your demand for heat inside the home increases, the ability of the heat pump to provide that heat efficiently decreases.

The Heat Pump and You

You should consider a heat pump, as well as other heat sources, if you are building or modernizing a home, replacing a worn heating system, or worried about ever-mounting fuel bills. And if you are planning to add air conditioning to your home, you should by all means explore the advantages of a heat pump over simple air conditioning.

Your first step, after reading this book, will be to discuss your situation with at least two heat pump dealers.

They will have questions as well as answers for you. This is a good time to determine whether a dealer is one you wish to work with. Will he give you the names of several of his customers so that you can chat with them? Does he try to sell you immediately, or does he explore your needs? Ask about his service program, and whether his staff members have more than on-the-job training. The model he handles may be the best, but the chances are good that you will be wanting service at some time in the future.

I discussed sales with one dealer who particularly impressed me. He makes a detailed study of a potential client's home. He finds out the area to be heated and cooled, the number of doors and windows, the heating bills of previous seasons, and many other details. He feeds all of this information into a computer programmed to give advice to the client, and at no cost.

"Our first recommendation may be for the homeowner to

tighten up his home, to add insulation, or plug up gaps that are leaking heat — and we don't even do that kind of work. But this may be the wisest spending, the best investment the homeowner can make,'' he told me.

When he recommends a heating system, it is not always a heat pump. "There are some situations in which a heat pump simply isn't the answer,'' the dealer explained. "Here's an example. Heat pumps require fairly large ducts to move the volume of air required, since the air usually is heated only to 100° F. If your house is a three-story, old house with poor insulation and small ducts, we probably won't recommend a heat pump. It simply would not be cost effective to install the large ducts you would need.''

Model recommended

If a heat pump is recommended, the homeowner is told what model and size unit to buy, its cost, its estimated operating expenses, and the anticipated payback period (the length of time it will take for the homeowner to get his money back).

The computer printout is given to him for checking and study, or to show to someone experienced in the heating/cooling field for advice.

The value of this computer check on your personal situation is that the computer is programmed to do much of the basic mathematics that will be described later in this book, do it quickly, and furnish you with results in a form that will be easily understood.

Is it a good sales gimmick? Of course. It's impossible not to be impressed with this little machine spewing out line after line of information about your house and its heating needs.

But can it be trusted? In most cases the answer must be yes. Manufacturers and dealers remember too well the bad name the heat pumps acquired when the emphasis was on sales, rather than service. They won't make that mistake again. And they realize they're in a market in which the companies that offer an inferior product and poor service will be edged out, despite the growing demand for heat pumps.

Visit installations

During the period your dealer or dealers are preparing recommendations for you, try to visit several homes with heat pumps. Most owners, if approached politely and given an opportunity to set a time when a visit will be convenient, are happy to talk about their heat pumps.

Try to select home situations similar to yours, houses of about the same size and age. Ask about heating and cooling bills before and after the heat pump was installed. See what their heating requirements are. If they want 75° F., there will be a big difference in the bills, even if the homes are identical. Small children drive up heating bills as they go in and out of the home, particularly if they are prone to leaving the door open for extended times.

When you return to your dealer, he will be ready with specific recommendations for you.

He may want to give you the big picture first. It's a dramatic one, when sales of heat pumps are described. Here's the residential sales picture:

	1970	1972	1974	1976	1978	1980
New heat pump sales	55,000	81,000	97,000	250,000	500,000	750,000 (est.)

About 70 percent of these sales are for new homes; 25 percent are for older homes, with many of these "add-on" units, replacing some other form of heating as the *primary* heat source, and the remaining 5 percent are replacement heat pumps.

It's time to talk costs with your dealer. His first figures may be higher than you expected. For a three-ton unit to heat and cool the average home, the cost will be about $3,500. That's a unit that will handle all of your heating and cooling needs. The figure includes installation costs.

There may be some extras. If your present system has small or no ductwork, the cost for installing this will be quoted. Another extra expense may be for electrical work. Many conventional houses have 100 to 150 amperes of power entering the house. That places a definite limit on the number of appliances that can be used. A heat pump will call for 60 to 80 amperes, which means your electrical system will have to be strengthened before the heat pump can be turned on.

The dealer may have better financial news for you. He may recommend an add-on heat pump, particularly if you live in a cold climate and have a forced-air heating system that is in good working condition.

This will cost about $1,000 less than the full heat pump unit. It's cheaper because it will not have the usual resistance system that is a part of the full unit.

With an add-on unit you will be using your heat pump for cooling, and for when the heat pump works the most efficiently — at less than extremely cold temperatures. The heat pump will be wired with your present unit so that the latter helps with the heating task at a time when it works most efficiently — at the coldest temperatures.

Here's a chart to show you comparative installed costs:

	HEAT PUMP	GAS FURNACE	ELECTRIC RESISTANCE	ELECTRIC FURNACE	PROPANE
Heating	$3,500	$1,000	$ 900	$1,200	$1,000
Cooling		$1,500	$1,400	$1,400	$1,500
Ductwork	$ 600	$ 600	$ 600	$ 600	$ 600
Chimney		$ 200			$ 100
Storage tank					$ 150
Total	$4,100	$3,300	$2,900	$3,200	$3,350

New Terms

By now your dealer may be tossing terms in your direction, terms you should understand. He may mention *single-package units* and *split-system units.* These terms refer to the number of parts the pump is divided into.

A single-package unit is just that — one box contains all of the elements of the heat pump, and is installed in the roof or wall of

Outdoor section of a Carrier split system air-to-air heat pump.

your home. Think of a single window air conditioner, used to cool one room. That's a single-package unit.

A split-system unit has two boxes or assemblies, one inside and the other outside the house. They're connected by cables and pipes through which the heat actually flows.

The dealer may next mention air-, water-, and earth-source options as well as solar-assisted heat pumps. In each case he is referring to the source of heat.

Air-source

Air-source heat pumps are the most common in the United States. They take heat from the outside air and deliver it to a hot air distribution system inside the house. The efficiency of these heat pumps will depend on your climate. Frequently they are not recommended in climates where winter temperatures often fall below 10° F. unless installed in conjunction with an alternate fuel. For them to be efficient in supplying heat, the difference in temperature between the outside air (heat source) and the end use must be as small as possible. If the difference is too great, the energy required to extract the heat from the source will be more than the heat derived. When a pre-set level of inefficiency is reached, the heat pump automatically switches on the more expensive electric resistance heating or an alternate fuel system.

Sources of Heat

TYPE HEAT SOURCE	AIR	WELL WATER	SURFACE WATER	EARTH	SOLAR
Suitability	good	excellent	varies	good	good
Availability	every-where	80% of U.S.	restricted	every-where	every-where
Capital cost of heat pump	lowest	medium	medium	medium	uncertain
Running costs	high	low	low	low-medium	can be lowest
Total cost of heat pump system	low	high if well is required to be drilled	medium	high	high
Stability of heat source/ sink	extreme	stable when large volumes of water are available	fairly constant	stable	extreme day night, seasonal
State of art	mature	mature	mature	growing	immature
Problems	frosting	calcification, second well may be necessary, well may run out	salt water corrosion	leaks hard to find, cost of earth coil high, difficult to install	high cost of solar panels, storage required for night/ clouds

Water-source

The water-source heat pump differs from the air-source unit in that it extracts heat from a water supply, quite often a well, but possibly a lake or river. The water goes through the unit, then is returned to the source or pumped back beneath the surface of the earth.

Water-source heat pumps have one big advantage. Throughout much of the United States, well water never falls below 50° F., making it an ideal heat source for heating and heat sink for cooling in summer. The accompanying map shows the approximate temperature in your area, and it should be noted that the temperature varies little from summer to winter.

Those considering use of well water should get an estimate of extra costs such as digging the well and installing the pump.

Earth-source

In most areas of the United States the temperature of the earth below the frost line rarely drops below 50° F. Earth coils must be placed at least one or two feet below that line. Laying the pipes is no problem in new construction when the earth-moving equipment is already at the site. However, in retrofit situations existing water lines and the septic system can be easily disrupted. In these systems an anti-freeze is circulated to absorb the heat.

Solar-assisted

Your dealer may not even mention this possibility because solar collectors, if used, are expensive, but I anticipate more and more installations will make use of the sun. These heat pump systems use the relatively low-level heat available in solar installations and boost it to levels for heating homes. Heat storage units are a part of these systems so that heat will be available day and night and during cloudy periods.

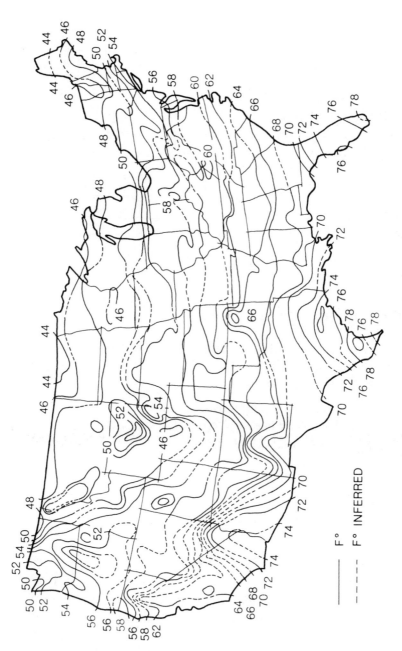

Ground water temperatures. Map courtesy of National Water Well Association.

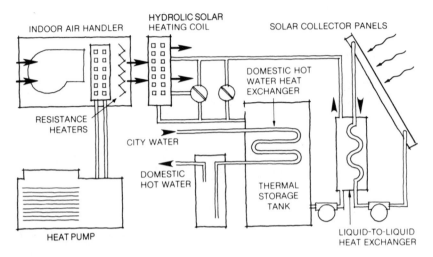

This solar-augmented heat pump system provides for the heat pump to back up the solar system, and illustrates versatility of heat pump systems.

Heat pump combinations

Those working with heat pumps also use terms to describe not only the source but the method of transporting heat in the house. There are four such combinations.

1. Air-to-air. By far the most common. Air is the source of the heat; the house has a system of hot-air ducts to transport the heat.
2. Air-to-water
3. Water-to-water
4. Water-to-air

These we will examine later in detail.

Efficiency

As you discuss heat pumps with your dealer, you'll notice that the term *efficiency* is used frequently. There are some basic things you should understand about efficiency.

When the costs are broken down for heat pumps, the most efficient units are those used in relatively warm climates where the demand for heat is not great, and the unit is called on to provide cooling during the summer.

The homeowner in cold climates can improve the efficiency of his *total* heating system by providing a source of supplemental heat that is as inexpensive as possible. Use of the heat pump as an add-on unit, as we explained earlier, is one method of doing this. For some areas, the source of supplemental heat might be a wood or coal stove. Most heat pumps provide their own supplemental heat, with a resistance heating unit built into the heat pump. For many homeowners this may not be the least expensive method of providing additional heat. Nor is it best for a utility system, since the utility would have to have a high peak-power level to meet the demand if it furnished electricity for a large number of heat pumps.

That heat pumps work most efficiently in more temperate climates points out the conflict in their operation. As the graph shows, when outdoor temperatures drop (and when you would like your heat pump to be at its best), the ability of the heat pump to furnish heat lessens. At the same time the demand for heat by the house increases. At some point the two lines cross. That point is called the *balance point,* when the heat pump can just barely provide enough heat for the home. This point varies with each installation, but generally it is between 18° and 28° F., and above in inefficient installations.

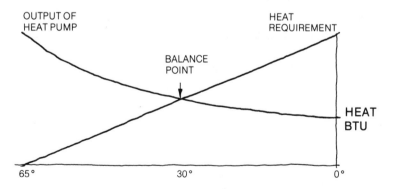

At balance point, heat pump provides just enough heat to meet needs of house.

The wise homeowner will try to make that balance point occur at as low a temperature as possible. He can do this in several ways. Some of them are:

1. Accustom yourself to living in a house heated to only 65° F.
2. Operate your heat pump in an energy-efficient home, one that is well insulated and doesn't leak heat — perhaps even gains part of its heat from a passive solar system.
3. Purchase a heat pump designed to operate in cold climates.

Measuring efficiency

If you are to understand the discussions with your dealer, you should learn the methods used to measure the efficiency of heat pumps.

We've all had a taste of measuring efficiency since gasoline occasionally has become scarce and always is expensive. We've learned to figure our car's mileage between fill-ups, and to divide that figure by the number of gallons of gasoline we've pumped into our tank, and found out the miles per gallon (mpg). The first measurement used with heat pumps is no more difficult than this, and tells us a similar story — how much we've gained in heat for what we've put into our heat pumps.

COP

For heat pumps it's called COP — coefficient of performance. It's a measurement of how much energy you put into a heat pump, and how much heat you get out of that energy. It illustrates the magic of heat pumps.

The formula looks like this: [Remember that one kilowatt-hour (1,000 watts) of electricity equals 3,413 Btu of heat.]

$$COP = \frac{\text{Btu of heat delivered to the home}}{\text{Btu of electric energy put into system}}$$

or

$$COP = \frac{\text{kw output}}{\text{kw input}}$$

Think of resistance heating, as from an electric heater. If one kilowatt hour of electricity (or 3,413 Btu of heat) is put into the heater, that same amount of heat is taken out. Thus the COP is one.

The amount of heat produced by the heat pump is *greater* than the Btu equivalent of electricity used to operate the pump. If the Btu equivalent of four kwh of heat were poured into a house, and only one kwh were used by the pump, the COP would be four — you would get four times as much heat from this electricity going into a heat pump rather than a resistance heater.

COP is one of the factors that determine the cost of operating a heat pump. The other is the cost of electricity. The accompanying chart gives an indication of comparative costs. The figures in the chart are for the number of cents you will pay for the equivalent in heat of one kilowatt-hour of electricity (3,413 Btu).

	H.P. COP 4	H.P. COP 3	H.P. COP 2	ELEC. HEAT	OIL FURNACE	GAS FURNACE
Elec. 2¢/kwh	0.5	0.7	1.0	2		
Elec. 4¢/kwh	1.0	1.3	2.0	4		
Elec. 6¢/kwh	1.5	2.0	3.0	6		
Oil $1/gal.					4.1	
Gas 56¢/therm						3.17

Comparative heating cost in cents per killowat-hour equivalent, assuming a 60 percent efficiency for oil and gas furnaces.

Estimating savings

The above table can be used to estimate roughly the annual savings from using a heat pump. For example, consider a house of 2,000 square feet, burning 1,000 gallons of oil annually at a cost of $1,000. Heat output equals 24,000 kwh. A heat pump with a COP of three in an area where the utility rates for 1 kwh is $0.04 will provide the heat equivalent of 24,000 kwh for $320. The use of the heat pump represents a savings of $680 annually, when compared with the burning of oil.

The Long Island (NY) Lighting Company in 1980 obtained similar results in a study. The company's computer analysis indicated that to heat an average home there, the costs were $811 for oil, $748 with electric baseboard heat, $455 with gas, and $425 with an electric heat pump.

After marveling at the obviously better job a heat pump can do, you might look back at the mathematics involved and ask: so what?

The "so what" is that engineers use the COP figures so that you and I can compare Brand A and Brand X to get some idea of which pump will operate most efficiently for us. The COP figures are obtained by laboratory tests, so can be compared.

As we've seen, the COP will vary as the outside temperatures fluctuate. The heat pump works harder the lower the temperature drops, so the COP decreases with a decrease in temperature.

High COP's are obtained when the difference in degrees between heat source and end use is the least. For this reason an excellent COP is found when the heat source is well water at 50° F. or more, when a solar system provides a "warm" source of heat, or simply when the outer air temperature is not below freezing.

For the same reason, the COP will be higher if those in a home are satisfied with temperatures of 65° F. or less, rather than above 70° F.

Just as mpg for cars is listed two ways, for highway and city driving, the COP is listed twice. The tests are for an inside temperature of 70° F. in both cases, but for outside temperatures of 47° and 17° F. A typical heat pump might be listed at 2.5 to 3 for 47° F., and 1.7 to 2 for 17° F.

Dealers have these figures for their own and competing models, and should provide them for you. Many are listed in the back of this book. You can also find them in directories published by the Air-Conditioning and Refrigeration Institute, 1815 N. Fort Myer Drive, Arlington, VA 22209.

EER

Heat pumps provide cooling as well as heating, and COP measures only heating. The EER (energy efficiency ratio) was devised to measure the efficiency of units operating in the cooling mode. It's much like the COP, as you'll see, and is the formula used in laboratories to measure the efficiency of both heat pumps and air conditioners.

$$EER = \frac{Btu/hour\ of\ heat\ rejection}{Watts\ of\ input}$$

An EER of eight or nine is good for most heat pumps.

Readers of a mathematical bent will quickly wonder why the COP and EER examples are so varied, why the COP figures are so much lower than the EER figures.

The answer is that the figures used to compute COP and EER are not the same. In the latter figure the heat rejection (which is comparable to the top line of the COP formula) is a measurement of Btu, while the lower line is *watts* rather than Btu input. Since there are 3,413 Btu in one *kilowatt-hour,* you must divide EER figures by 3.413 to get a figure that can be compared with COP.

SPF

We now know how to figure the efficiency of the heat pump in its heating and cooling phases. We need one more measurement. This is called *Seasonal Performance Factor.* It is a measurement of the efficiency of a heat pump over a season as compared with electrical heat.

$$SPF = \frac{Btu\ from\ heat\ pump\ +\ Btu\ from\ auxiliary\ resistance\ heat}{3,413\ (kwh\ to\ run\ heat\ pump\ +\ kwh\ for\ auxiliary\ heat)}$$

If the SPF figures out to be two, as an example, the system would use only half as much electricity as a resistance system. SPF is computed in the laboratory for individual heat pump models. And, since it is so directly linked with seasonal temperatures, it is also used to predict heat pump operation throughout the country, as the following map shows.

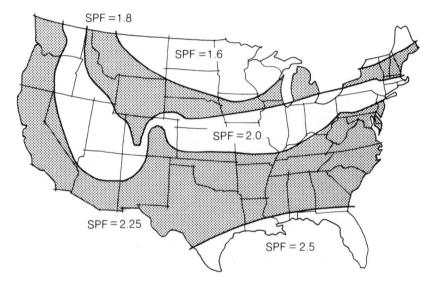

Seasonal Performance Factor figures are closely linked with temperatures.

LOCATION	DESIGN TEMP. OUTSIDE	DESIGN TEMP. INSIDE	RESISTANCE HEATING KWH	RESISTANCE HEATING COST	HEAT PUMP KWH	HEAT PUMP COST	ENERGY SAVING PERCENT	SPF
Atlanta	2 F.	70 F.	4,808	$240	2,164	$108	55	2.22
Boston	7 F.	70 F.	7,291	$364	4,401	$220	39	1.66
Chicago	12 F.	70 F.	7,243	$362	4,913	$245	32	1.47
Dallas	7 F.	70 F.	4,217	$211	1,966	$ 98	53	2.14
Houston	17 F.	70 F.	2,951	$148	1,237	$ 62	58	2.38
Kansas City	2 F.	70 F.	6,697	$335	4,291	$215	35	1.56
Los Angeles	32 F.	70 F.	7,349	$367	2,811	$140	61	2.61
Miami	37 F.	70 F.	1,235	$ 62	472	$ 24	61	2.61
Minneapolis	27 F.	70 F.	7,489	$375	6,120	$306	18	1.22
Nashville	7 F.	70 F.	4,876	$243	2,579	$128	47	1.89
New Orleans	17 F.	70 F.	2,786	$139	1,110	$ 55	60	2.51
New York	2 F.	70 F.	7,145	$715	3,840	$384	46	1.86
Phoenix	27 F.	70 F.	4,595	$230	1,839	$ 92	60	2.50
San Antonio	7 F.	70 F.	2,891	$145	1,255	$ 63	56	2.30
Seattle	12 F.	70 F.	10,571	$529	4,712	$236	55	2.24
Tampa	27 F.	70 F.	2,559	$128	996	$ 50	61	2.57
Washington, DC	12 F.	70 F.	7,365	$368	3,750	$188	49	1.96

SPF rating and cost comparison of air-source heat pumps and resistance heating.

Warranty

By the time you've decided to buy a particular model of heat pump from a particular firm, you should have talked with its representatives, seen some of their installations, and talked with persons for whom they have done work. There's also one more step you should take and that is to read the warranty the firm you've selected will provide.

A typical factory warranty covers the replacement of any defective parts for one year, and the compressor (often the most vulnerable part) for five years. The dealer will furnish labor for any such work free of charge to you, for the first year's work. Thus you will pay for all parts except compressor parts after the first year.

Many reputable manufacturers also offer a service contract. This usually costs about $50 a year, and covers both parts and labor. Some even offer inspections every six months.

If you have read this chapter carefully, you're ready to discuss heat pumps with your dealers, to ask intelligent questions, and make use of the answers. But you'll find that in many areas, a little more understanding would be a great help. So read on — before you make the decision on whether there's a heat pump in your future.

The Broad Array
of Heat Pumps

Take a walk with me through a warehouse, a warehouse that exists only in our minds. In it are all the heat pumps that have been manufactured or experimented with, or even thought about. You'll notice several classifications. These include size, of course, as well as source of heat, how they spread that heat in your home, and several other categories.

Air-to-Air

Let's take a look first at the air-to-air heat pump. It's the most common and the least expensive (although it may not be the cheapest to operate).

Air-to-air means heat is taken from the air (that's the first "air") and is circulated in the house through a forced hot air system.

You have your choice of two air-to-air systems.

1. Single-package system. All of the components are packaged in one unit, which is mounted on the side or roof of a house, and ducted through a wall, the roof, or the foundation. In general, efficiency of these units is slightly lower than the other system.

Here's a Fedders single-package system, with all components in one unit.

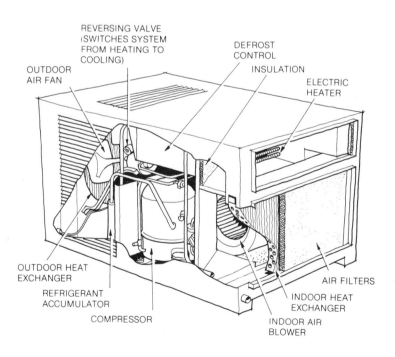

Components of a single-package system. This unit could be placed in a window.

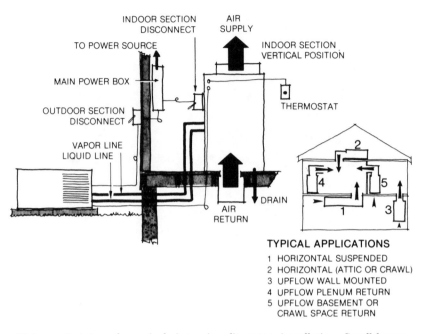

Wiring and piping of a typical air-to-air split-system installation. Small house shows variety of ways the inside unit can be installed.

2. And that is the split system. It has two units, with one unit inside the house, one outside, and the two linked by pipes and cables. The split system usually is quieter than the single-package unit.

The outdoor and indoor heat exchange coils are similar in air-to-air heat pumps. Both are made of finned copper pipe through which the refrigerant flows. Both coils have air-circulating fans to accelerate the exchange of heat between air and refrigerant.

Air-to-air heat pumps have four-way valves. These determine whether the unit is providing heat for the house or taking heat from it. The refrigerant passes through the evaporator where it absorbs heat from the air outside, and is changed by that heat to vapor. It moves next into the compressor where it is compressed to a higher pressure and temperature. This vapor then passes through the condenser, which is the indoor coil, gives up its heat to the air

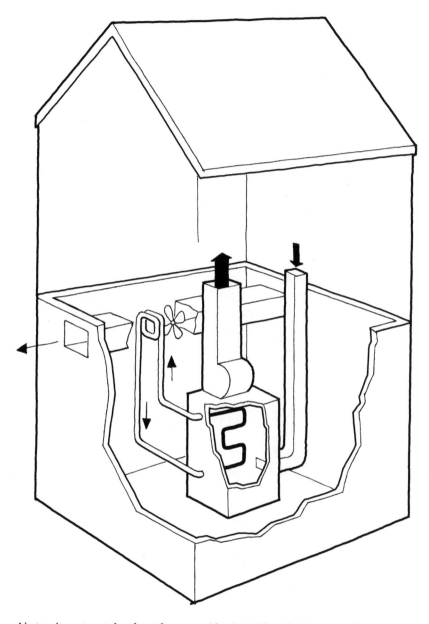

Air-to-air system takes heat from outside air (with unit shown inside house in this sketch), and moves it into larger unit, where flow of air removes it.

circulated past the coil, and changes back to a liquid. It goes through the expansion valve, expands, and at the same time drops in temperature. The cycle is completed as it moves again into the evaporator, ready to pick up heat once more.

During the cooling cycle the flow of refrigerant is reversed by that four-way valve. When on the cooling cycle, the liquid turns to vapor in the indoor coil as it absorbs heat. The refrigerant is then compressed to a higher pressure and temperature in the compressor before it passes to the outdoor coil where it condenses into a liquid, giving up its heat to the outdoor air. The liquid then passes through the expansion valve, expands, and is ready to take more heat out of the home.

The greatest single advantage of air-to-air heat pumps is that air is so convenient both as a source of heat and as a means of distribution. Outside the home only the coil and its fan are needed. Ducting is needed inside the home to distribute the heat. Because the heat being distributed is often of a lower temperature than that from a furnace, more air must be circulated, and larger ducting is required. This can be an additional expense in homes where a heat pump is replacing a furnace system, since the old ducting may not be large enough.

Air-to-Water

Air-to-water heat pumps use two different types of heat exchangers. The outside heat source coil is a finned copper tube, through which the refrigerant passes. Air is blown over the tube by a fan to allow a heat exchange to take place between the air and the refrigerant in either the heating or cooling mode, just as in the air-to-air system.

The refrigerant-to-water heat exchanger indoors uses water rather than air as the heating and cooling medium. These heat exchangers are either counter-rotating tubes wound around one another and insulated to allow good heat transfer, or else the refrigerant passes through a section of tube placed in an open water channel.

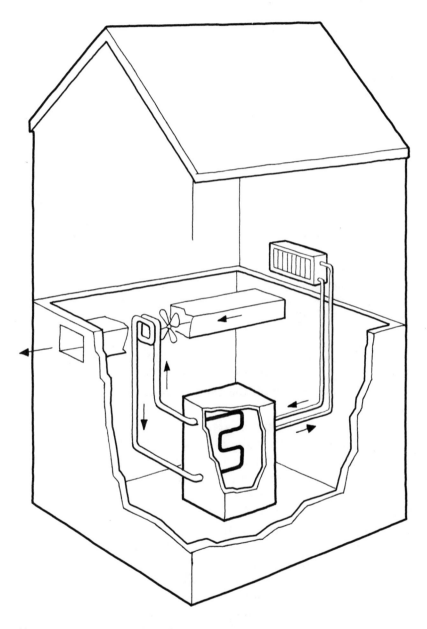

Air-to-water system uses heat from outside. Warmed refrigerant flows through pipes to heat exchanger in boiler. Heated water flows into radiators.

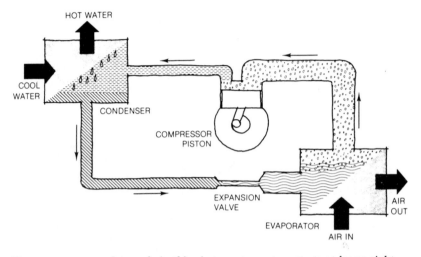

The vapor compression cycle in this air-to-water system starts at lower right. Liquid refrigerant picks up heat and changes into vapor. Heat level is raised in compressor. Heat is discharged in condenser, and cooled refrigerant flows through expansion valve back to evaporator.

As with the air-to-air heat pumps, the air-to-water heat exchange coils, both indoors and out, act as condenser and evaporator, depending on the cycle of operation — be it heating or cooling.

Air-to-water heat pumps are rare in the United States. In Europe, hydronic (wet radiator) heating is common. Hydronic systems heat by radiation, which you may prefer to the more drafty, hot-air systems. In both this country and abroad, underfloor hydronic heat is an increasingly popular heating alternative for new home builders.

A major difficulty with air-to-water heat pumps is that heat at a higher temperature is required in the house. To supply heat effectively, hydronic systems need to supply hot water to the radiators at between 120° F. and 140° F., whereas air used as a heating medium only needs to be heated to between 85° F. and 110° F. This very obvious disadvantage of using water to circulate the heat within the house can be in part eliminated by increasing the heating radiator surface.

Water-to-Air

There, down under your feet, is the world's largest storehouse of heat. It's the ground water system, ready to provide a source of heat that remains constant, summer and winter. In many areas it is 50° F. or higher.

Using it with a heat pump will be more expensive initially. You'll need a well, casings for that well, and a pump to bring the heated water up so that it can be used. You may also need a dry well so that the water you've used can be put back into the ground.

That initial cost may be paid back to you in savings in six or eight years.

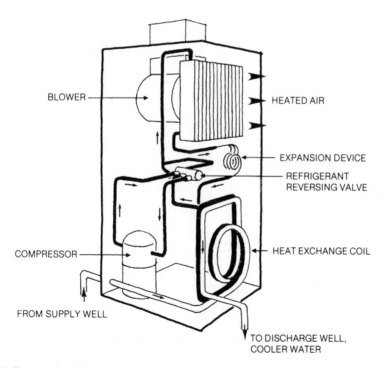

Well water furnishes heat, warming refrigerant in heat exchanger coil. Refrigerant, compressed, flows to top of unit, where flow of air removes heat.

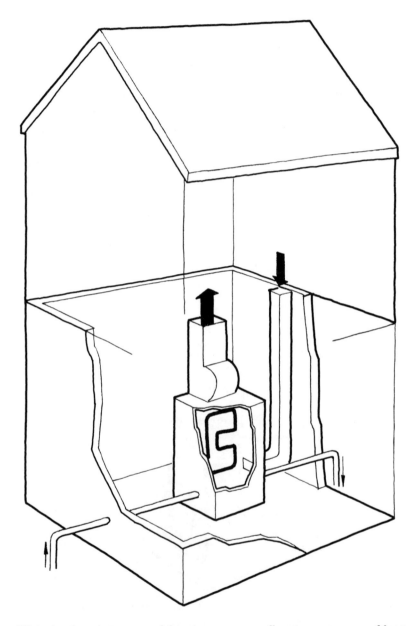

Water-to-air system can use lake, river, sea, or well water as a source of heat. Air moves down into plenum, is warmed, and flows back into living area.

The Milwaukee, Wisconsin, *Sentinal* interviewed one water-to-air heat pump owner who bragged of winter heating bills running no more than $80 a month with the thermostat set at 75° F., while a neighbor braved it at 65° F. with a gas heating system, and paid $385 a month.

And that isn't an isolated case. The federal Environmental Protection Agency commissioned a survey in 1976 which concluded that such heat pumps could be used in "over 75 percent of the continental United States. A majority of the thirteen million homes supplied by ground water have enough water to meet both domestic and heat pump demands. The 500,000 new homes built annually and supplied by domestic wells could reap significant energy savings (30 to 60 percent) through the use of ground water heat pumps." EPA and the Department of Energy are studying both the possibility of nationwide use of this system as well as any environmental impact it might have.

This system is in common use in some parts of the nation already, particularly in Florida and the Southeast where there is an abundance of warm ground water, making possible high COP's in the operation of the heat pumps. Water as a heat source is, in the depth of winter, usually higher in temperature than air. This means water-to-air heat pumps are commonly 25 percent-plus more efficient than air-to-air units. Air temperature frequently falls below 32° F. in winter in most parts of the country, but ground water never freezes, at least not below the frost line. This affects heat pump efficiencies; the lower the temperature difference between heat source and heating needs, the greater the efficiency of the pump.

The homeowner considering a water-to-air heat pump should be prepared for a certain amount of red tape with the government. The digging of wells is legal in all states, but in most states a permit is required, and in areas within many of the states, the digging of a well is prohibited. For example, in some communities where there is a public water system, digging a well is prohibited to avoid the possibility of contaminating the drinking water. In such places, it may be possible to get a variance when the proposed use of the water is explained.

If you have a well system, you may be able to add a water-to-air heat pump at little additional cost, provided your well can provide the additional water required. Estimate about three gallons per minute for each ton of heat pump capacity — but check out the requirement for the heat pump you're considering. It may demand more — or less — than this.

Water-to-Water

Water-to-water heat pumps use two water-to-refrigerant heat exchangers.

In the heating mode, heat is extracted from the water source — well water, lakes, or the sea — and is passed on by the refrigerant to the water used for heating. The reverse takes place in the cooling cycle.

Such heat pumps have the same disadvantage as those just described for air-to-water systems — the need to heat water to a higher level for satisfactory heating within the home.

Add-On Heat Pumps

An add-on heat pump can be hooked to an existing forced-air heating system. Any heat pump can be used as an add-on unit, but some manufacturers have designed special control and installation systems enabling them to market "add-on," sometimes called "hybrid," heat pumps.

Add-on units are designed so that when heat-source temperatures fall so low in winter that the cost of operating the heat pump rises above that of heating with oil, LP gas, wood, or electricity, the original heat source is automatically fired. This occurs at the *balance point* — the temperature setting above which the heat pump will operate alone and below which the existing system is switched on. The control box allows for an adjustable balance point and automatically switches from one heat source to another.

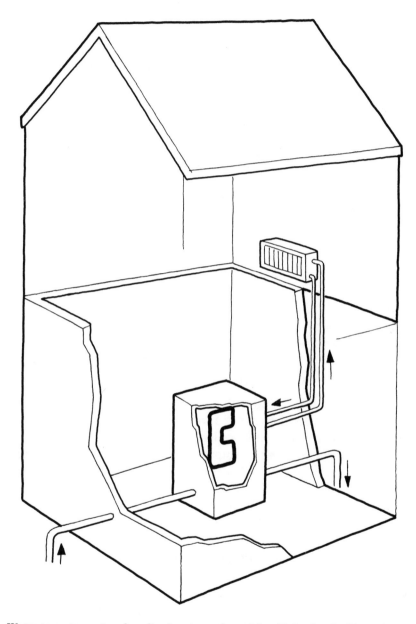

Water-to-water system has disadvantage of requiring higher level of heat than air system, above 120° F. rather than above 85° F. for air system.

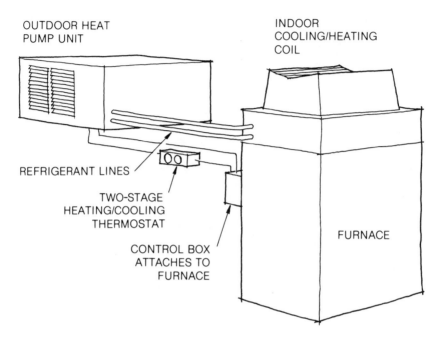

OUTDOOR HEAT
PUMP UNIT

INDOOR
COOLING/HEATING
COIL

REFRIGERANT LINES

TWO-STAGE
HEATING/COOLING
THERMOSTAT

CONTROL BOX
ATTACHES TO
FURNACE

FURNACE

The Lennox Fuelmaster hybrid heat pump can be attached to hot air furnace.
Controls can be set to operate system that is more efficient for conditions.

The BDP Company control package that makes the changeover at the balance point is called the *Energy-Minder Control* and is designed for use with BDP split-system heat pumps.

The proper setting for the balance point can be found by obtaining local costs for gas, oil, electricity, or other alternative fuel. Contact your local fuel supplier or utility for cost figures to use in making estimates.

Manufacturers of add-on heat pumps have developed graphs to enable local installers to calculate the balance point. To illustrate the process, we will use a graph produced by BDP Company.

The graph is for LP gas. Let us suppose one can buy electricity for 5¢ a kwh and LP gas for 50¢ a gallon.

Follow these two lines on the graph to the left until they cross. Next, from the intersection of these two lines, follow the line to the bottom of the graph. The outdoor temperature, corresponding to

FUEL COST BREAK-EVEN GRAPH
HEAT PUMP VERSUS L P FURNACE*

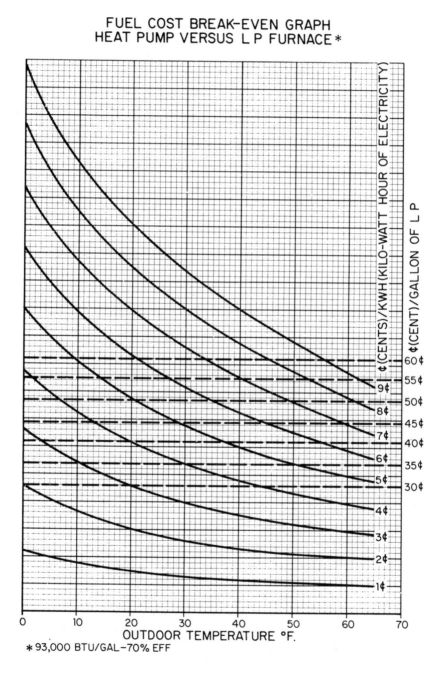

* 93,000 BTU/GAL–70% EFF

the vertical line, is 20° F. When the temperature falls below 20° F. the back-up oil, gas, or whatever should be turned on.

There is a simple, although crude, system that homeowners can use to calculate the potential cost savings. Look back over your past fuel bills to determine how many gallons a year you burned.

Say the figure is 1,000 gallons at an estimated 1980 cost of $1.00 a gallon, or a yearly cost of $1,000. This figure of 1,000 gallons needs to be converted into kwh's so that the cost of heating with electricity, and therefore a heat pump, can be calculated.

$$\frac{1,000 \text{ (gal.)} \times 130,000 \text{ (Btu/gal.)} \times 0.6 \text{ (Efficiency)}}{3413 \text{ (Btu/kwh)}} = \frac{22,853}{\text{kwh equivalent}}$$

At 5¢ a kwh this would cost $1,142 a year for electric resistance heating.

To find out how many kwh would be required to get the same heat annually with a heat pump, you need to know the type of heat pump (air or water source); the type of installation (solar- or environmentally-assisted or neither); the COP of the heat pump; and, more important, the likely SPF. An indication of the SPF of air-to-air pumps, installed without regard to the wind chill factor, solar gain, or any other environmental factor is shown on page 33. These are among the lowest possible SPF's. The opposite extreme, that of a high-efficiency water heat pump using a high temperature (50° to 70° F.) water source, can have a SPF as high as four.

To return to our example of 22,853 kwh (the equivalent of 1,000 gal. of fuel oil), this same amount of heat would cost $571 if a heat pump with a SPF of two were used and $381 if the heat pump had a SPF of three (possible with some water-source and solar-oriented air-source heat pumps). The COP figures included in the Manufacturers' Index at the back of this book give an indication of the SPF's, but the best way to get the SPF is to ask the dealer or manufacturer of the unit(s) you are most interested in.

If one is using LP gas, natural gas, or wood, the kwh equivalent can be ascertained by finding out how many gallons of LP gas, hundreds of cubic feet of natural gas, or cords of wood it takes to heat your home annually and converting the figures as follows:

LP gas: $\dfrac{\text{X (gal). } \times \text{ 93,000 (Btu/gal.) } \times \text{ 0.6 (Eff.)}}{\text{3,413 (Btu/kwh)}} = \text{kwh}$

Gas: $\dfrac{\text{X (CCF) } \times \text{ 100,000 (Btu/CCF) } \times \text{ 0.6 (Eff.)}}{\text{3,413 (Btu/kwh)}} = \text{kwh}$

Wood: $\dfrac{\text{X (cords) } \times \text{ X (Btu/cord) } \times \text{ 0.6 (Eff.)}}{\text{3,413 (Btu/kwh)}} = \text{kwh}$

Any other fuel: $\dfrac{\text{Fuel quantity } \times \text{ Btu per unit } \times \text{ burner efficiency}}{\text{3,413 (Btu/kwh)}} = \text{kwh}$

The furnace efficiency of 60 percent is average; it will differ form unit to unit.

By comparing the cost savings from using an add-on unit, and considering initial costs, you can decide whether heat pumps offer you savings' possibilities.

One advantage of having an add-on unit is that unless it is used to replace resistance heating, it gives the user multi-fuel capability. In an age of uncertain fuel supplies, this is good insurance. The best energy source for the home is many sources, such as some combination of wood, oil, gas, coal, and electricity.

Improved home insulation is often required to get the greatest efficiency from any heating system. Minimum insulation of the following thermal resistance values is recommended: ceilings, R-19; walls R-11; floors, R-11. Insulation of R-19 in walls and R-29 in ceilings will, of course, be even more efficient.

A problem with add-on units is that in many homes the duct system will have to be enlarged to handle the greater air volume needed with the relatively cooler air of the heat pump. The duct system should carry from 400 to 500 cubic feet per minute per 1,200 Btuh of ARI-rated (Air-Conditioning and Refrigeration Institute) cooling capacity of the heat pump.

If the existing air ducts are too small, there are two options:

1. Retrofitting a house with new ductwork. In some homes this may be relatively inexpensive, an average cost might be $3,000, and in some homes it would be prohibitively expen-

KVCC   KALAMAZOO VALLEY
COMMUNITY COLLEGE
LIBRARY

sive. If your bill for retrofitting were to cost $3,000, you might explore spending a like amount for insulation, thus enabling you to buy a smaller heat pump, one that would match the size of the existing ductwork. Tax credits, available for insulation but not for ductwork, may improve the attractiveness of this idea.

2. If the house is well insulated and the existing ductwork is still inadequate for the recommended heat pump, consider using a smaller heat pump to provide background heating, then adding an auxiliary heating system such as a wood or coal stove to meet the extreme demand during the coldest winter days.

To estimate the amortization of the add-on heat pump and installation, divide the total cost by the annual savings. Thus if the total cost of the unit is $3,500, and you save $500 annually by using it, the payback period is seven years.

When to buy

If you have an air conditioner and forced-air heating, the best time to buy an add-on heat pump is when the air conditioner needs to be replaced. The heat pump will cost a little extra, but the ducting for the air-conditioning unit is generally fine for a heat pump. And there is the added bonus of dual-fuel capability which makes good sense when there's always the possibility of fuel shortages of one kind or another.

An add-on heat pump may often be cost-effective when linked with an oil furnace; in the same situation a heat pump with the conventional back-up of resistance heating may not prove to be cost-effective.

Manufacturers who offer add-on heat pumps include:

Addison	Air-source
BDP Company	Air-source
General Electric	Air-source
Vanguard	Water-source
The Williamson Company	Air-source
York	Air-source

Heat-Only Heat Pumps

Heat-only heat pumps are potentially as great a wave of the future as cooling-only heat pumps (air conditioners) have been in the past. The prime market for heat-only heat pumps is where there is no need for air conditioning or where the need is so little that it can be satisfied by opening windows to create cross-ventilation.

I know of only one American manufacturer of heat-only heat pumps and that is Janitrol, owned by SJC Corp., which makes the Wattsaver. The Wattsaver is an air-to-air unit with a COP rating of 3.2 at 47° F. and a COP of 1.9 at 17° F. (See Manufacturers' Index under Janitrol.)

The Wattsaver is 10 to 20 percent more efficient than most heat pumps for two reasons:

1. The compressor and controls, both creating heat, are placed with the indoor coil. Thus this heat, which would be lost if they were outside, is saved and used effectively.
2. The heat-only unit can be designed specifically for heating. There is some compromise in the design of a unit that both

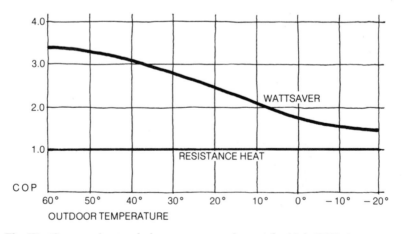

The WattSaver, a heat-only heat pump, can boast of a high COP, in part because it is designed only for heat, not to both heat and cool.

heats and cools, a compromise not made with the heating-only unit. The number of valves and electrical controls and the amount of wiring can be reduced with the heat-only unit. This increases efficiency and reduces costs.

In Europe, where summer cooling has never quite caught on, heat-only heat pumps are common. In England the Eastwood Heating Developments Ltd. (Portland Road, Shirebrook, Mansfield, Notts, England) has developed the Eastwood 80 Series of heat-only heat pumps with an average COP of 3.5 during the English heating season.

The evaporator coil of this heat pump is suspended from the roof trusses in the attic. With this placement the evaporator can gain heat from the winter sun via the roof which acts as a low temperature solar collector. The evaporator also benefits from the higher temperature in the attic due to heat loss from the house. Attic placement further reduces the wind chill factor.

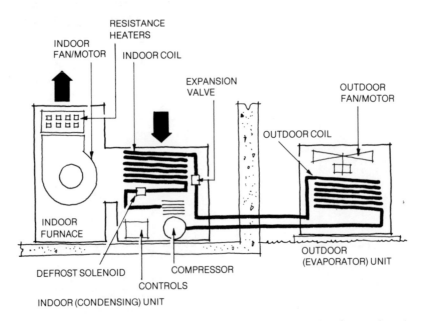

This WattSaver is linked to a forced-air furnace. Because it is a heat-only unit, it does not have the capability of cooling as well as heating the home.

Like the Wattsaver, the compressor and controls of the Eastwood are placed next to the condenser so that waste heat from them is used for house heating. The condenser and compressor can be placed in the garage or wherever convenient. The heat pump, which includes an auxiliary resistance heater for back-up, was offered in 1979 for 850 British pounds ($2,000 approximately).

High-efficiency heat-only heat pumps have great potential in domestic and commercial applications. A heat-absorbing attic is an ideal spot for the evaporator, provided that there is sufficient air flow. Attic placement also may eliminate condensation, an added bonus.

Large Scale Use

Advanced heat pump technology in industry, agriculture, shipping, and commerce is an obvious area of expansion for the heat pump industry. During the harsh winter of 1976-77 some factories closed as scarce gas supplies were shifted to top-priority users. Many firms were not prepared for the extended reductions in gas, and began to look for alternatives. Heat pumps are such an alternative either as a standby system or for fulltime use.

Agriculture

Agriculture, the food industry, and greenhouses are fertile fields for the utilization of heat pumps. Greenhouses are particularly energy-intensive because they lose heat so quickly. However, for much of the year the required temperature increase is relatively low. This makes a high COP possible.

Ken Hoyt of Solar Tech Corp. was the first person in America to design and install a water-to-air heat pump system for greenhouse heating. The pilot system is a Klupenger's greenhouse in Aurora, Oregon. According to the owner of the nursery, the system has reduced heating costs by 82 percent over the previous costs of using propane.

The water source for the heat pump is an on-site well with water at a constant 53° F., a temperature for which his heat pump was custom designed. The unit sends 65° F. air to the greenhouse. The low 12° F. temperature rise required for heating is perfect for high efficiency.

The greenhouses comprise 18,000 square feet and contain 170,000 cubic feet. The fans in the system maintain an air circulation of 37,500 cubic feet a minute, twenty-four hours a day, eliminating both high and low temperatures to optimize growth. Ten 100-foot plastic ducts provide air circulation.

The potential for this type of system, not just in the United States but all over the world, is great. It can be used for many agricultural and food applications, such as poultry farms, dairies, and grain drying. Hoyt claims his system cools cows' milk from 97° F. down to 37° F. and the 60° F. differential is used to produce hot water up to 180° F.

An interesting comparison to the Oregon greenhouse installation is a similar one installed by Escher Wyss of Zurich, for a market-gardener in Switzerland. The water-source temperature is between 50° F. and 54° F. and is available at a depth of twenty-

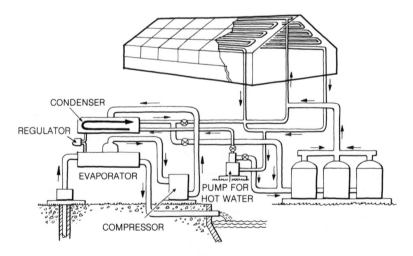

A water-to-water heat pump installation in a Swiss commercial greenhouse.

seven feet. About 110 gallons per minute are pumped through the evaporator and its temperature is reduced 7° F. by heat transfer. Hot water is delivered from the condenser at 120° F. to heat the greenhouse.

The heat pump replaces electric resistance heating and three coal boilers. In the year before the heat pump installation, 196,240 kwh of electricity were used. With the heat pump, this was reduced to 65,000 kwh and no coal was used as back-up. The system has a minimum COP of three.

Industry

The perfect client for a water-to-water heat pump is the Liffey Woolen Mills in Dublin, Ireland. The mills generate 90 kw of hydro-electricity from the river Liffey, but depend on oil to supply an additional 300 kw. Had the Liffey Woolen Mills installed a water-to-water heat pump with a COP of 3.3 (well within the bounds of possibility), their only dependence would have been on the river, a far more reliable source of energy than the OPEC countries.

Not every industry has water power available, but every industry that uses a heat process has the potential for energy savings from the application of heat pumps.

Dunham-Bush and Westinghouse are both active in manufacturing industrial heat pumps which can be used in the following areas: food processing, textile products, lumber, paper mills, chemical and petroleum production, metal and machinery, as well as transportation.

Westinghouse manufactures the Templifier, which has a solar option. The COP of the Templifier ranges from 2.5 to 6, depending on the application. It is used in industry wherever there is "free" warm waste water.

The first application of the Templifier was at Westinghouse's transformer plant in Muncie, Indiana. The heat-source water temperature is between 70° F. and 85° F. It is taken from a cooling tower serving a battery of welders. The heat is upgraded to 140° F.–170° F. to supply a degreaser and pre-painting treatment.

Even when near freezing, this river in Switzerland provides enough heat to warm the Zurich swimming baths, by use of a water-to-water heat pump.

Three Templifiers at Caterpillar's tractor plant in East Peoria, Illinois, soak up rejected heat from various pieces of process equipment and boost the temperature to 172° F. for use in washing tractor parts.

At the Wolverine Division of Universal Oil Products in Decatur, Alabama, a Templifier takes waste water from the cooling tower at 80° F. to 100° F. and boosts it to 170° F. The hot water is used to heat a hot soap process, other process tanks, washroom hot water, and the locker room. The system eliminates the need to use gas for these processes and so conserves gas for essential brazing needs. The Templifier, which has a COP of 3.4, is expected to pay for itself in less than a year and a half.

Shipping

The shipping industry, which floats in a vast heat reserve, could conserve oil by using water-to-water or air heat pumps for heating, cooling, and hot water supplies. Where most shipping occurs, sea water temperature rarely falls below 50° F., which is an ideal temperature for heat pumps. COP's of between three and five are possible. Direct mechanical power could be taken from the ship's engine to drive the heat pump.

Commercial

One popular type of commercial heat pump system is a centralized system using a closed water loop and numerous individual water-to-air heat pumps for separate rooms or working areas. The closed water loop acts as an energy reserve. It is kept at temperatures ranging from 60° F. to 100° F. from which individual units can heat or cool separate areas of the building. For example, a hotel may need cooling in the kitchens and service areas, heating in conference and dining rooms, and cooling in south-facing bedrooms. The various heating and cooling needs can be effectively served by one loop with the waste heat from the kitchens heating the dining room and so forth.

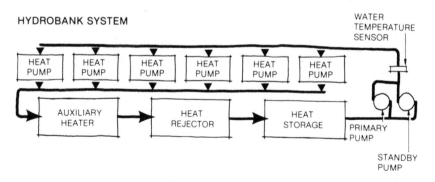

This large, centralized system, with heat pumps for individual rooms, can provide both heating and cooling for various areas of a building.

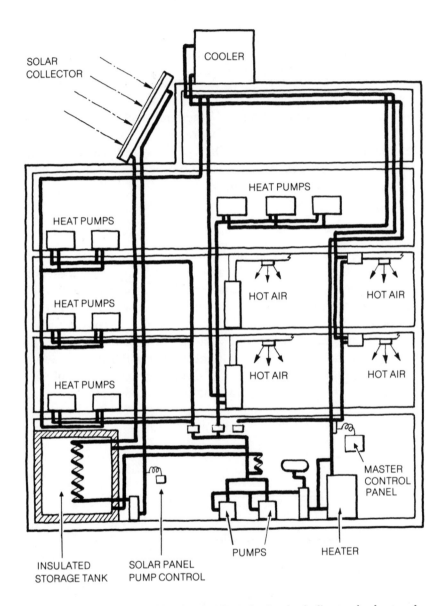

A system such as this offers the possiblity of using both direct solar heat and waste heat within an industrial building to both heat and cool the structure. It is particularly useful when heat may be required in one section of the building, cooling in another.

Manufacturers claim this type of system can reduce heating costs by about two-thirds for shopping centers, schools, supermarkets, hotels, and entertainment complexes. Where cooling is necessary, the waste energy can be used to heat water. This is exactly what the Hilton Hotel in Jacksonville, Florida, does. A Templifier heats service water to 125° F., using waste heat from the closed loop as the heat source, and does so at a COP of 4.7.

Another complex using the closed loop heat pump design is the Cincinnati Coliseum. Heat from the coliseum rink heats the rest of the complex, with the loss of heat freezing the ice. Only rarely must extra energy be purchased for supplemental heating.

Another potential customer for closed loop heat pumps is the movie theater. Many theaters require heating, then cooling as the movie nears its end.

Ideal for airports

Airports are also ideal for closed loop heating and cooling. Pitkin County Air Terminal in Aspen, Colorado, is of a passive solar design. Most of its winter heating and sumer cooling is supplied by bead walls on the staggered south walls and insulated skylids on the roof. The solar gain from these sources is absorbed by a thermal storage mass of solid interior walls and floors. The solar gain in summer, however, is more than desirable, as are the highs and lows in tempeature. A small closed loop heat pump system would effectively solve the problem.

A potential market for new heat pump installations in existing industrial, commercial, and agricultural plants awaits the enterprising manufacturer, energy engineer, and installation expert. Manufacturers of large-scale heat pumps include:

American Air Filter
Carrier Corporation
Command-Aire Corporation
Dunham-Bush
Friedrich

General Electric
Mammoth
Westinghouse
York

Desuperheater

A desuperheater is a device linked with a heat pump to heat water for a domestic water supply. The unit is used commonly to take heat from the heat pump during its cooling mode, when the heat pump is being used to get rid of unwanted heat. In the heating mode, desuperheaters can make a small contribution to hot water needs by removing heat from the "superheated" refrigerant as it leaves the compressor on its way to the condenser.

Depending on location and heat pump use, desuperheaters can provide from 30 to 80 percent of all hot water needs, approximately the same as solar panels. The difference between the two is the cost. A desuperheater, installed with a domestic hot water tank, costs less than $500, compared to the $2,500 for solar collectors with the same energy output.

Desuperheaters pay back their costs in four years or less for the average home; for the growing number of commercial installations using heat from refrigeration in restaurants and hotels, the payback period can be as short as one year.

Advantages

Advantages of desuperheaters are:

1. *No running costs.* Desuperheaters can provide a substantial part of the hot water requirement for free. Because the refrigerant flows around the heat pump anyway, there is no additional cost in removing the extra "superheat." A small water pump is required to circulate the hot water in the heat exchanger — at a cost no greater than that of a similar pump for a solar system — or this cost can be avoided by using a thermosyphoning system. Compared with solar, desuperheaters cost less initially; cost nothing to run; and use far fewer natural resources (copper), yet provide the same benefits.

2. *Improved heat pump efficiency.* Because desuperheaters effectively increase condenser surface areas by 25 to 30 percent, they serve to increase overall heat pump efficiency. As a result, the load on the compressor is reduced by 10 to 15 percent. Desuperheaters reduce heat pump running costs and thereby increase the COP.

How desuperheaters work

The refrigerant absorbs heat in the evaporator, and the gas is compressed to temperatures as high as 250° F. or more. From the compressor, the "superheated" gas is forced on to the condenser where it gives up its heat. In an ordinary air-to-air domestic heat pump, the air passing the condenser is heated to 110° F. or thereabouts. Clearly, not all of the 140° F. temperature difference is used in the condenser to heat the air. Some 30 percent of the heat in the superheated refrigerant is wasted in simply reducing the temperature to the point where it will condense effectively. It is this 25 to 30 percent of "extra heat" that the desuperheater renders useful.

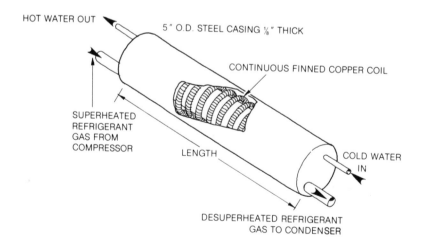

Desuperheater is simple in design, costs nothing to operate, and provides hot water.

Desuperheaters are simple refrigerant-to-water heat exchangers. They are installed between the compressor and condenser of heat pumps or any other refrigeration equipment. Just replace a section of the discharge line between compressor and condenser with the heat exchanger. The water is than fed either into a preheat tank or directly into the main tank, depending on use. The heat exchanger is designed to pick up a maximum of 25 percent of the rated load of the condenser.

Cost-effective

A coffee house in Corona del Mar, California, has desuperheaters fitted to two of its four compressor refrigerators. The energy saved is used to preheat washing-up water. The costs are as follows:

Solar Research Desuperheaters, part 5910	$42.00
5911	$58.00
25 gal. preheat tank (used copper tank with burnt-out electric element)	$25.00
Copper piping	$25.00
Freon 12 refrigerant	$15.00
Labor, one person — one day	$160.00
Total	$325.00

The $325 was quickly paid back in reduced gas bills, and since the system is thermosyphon, there are no capital or continuing costs involved for a water pump.

Manufacturers

Packaged desuperheaters are available from most manufacturers. For example, Friedrich manufactures the "Hot Water Generator," Carrier makes the "Hot Shot," and General Electric makes the "Hot Water Bank." Solar Research, 525 North Fifth St., Brighton, MI 48116, manufactures six desuperheating kits ranging in refrigeration compressor horsepower from one to ten, and in price from $42 to $206.

Marvair manufactures a $300 desuperheater called the "Heat Saver." Tests using a three-ton heat pump showed a recovery of approximately 60 percent of the heat removed from the home during the cooling cycle. This is equal to a six-kw heating element at no operating cost to the homeowner other than the small cost of powering a tiny pump.

During colder months when the unit is operating above the balance point, the Heat Saver uses a proportion of the heat absorbed by the heat pump from outdoor air to heat water used in the home. When the unit operates below the balance point, the outdoor thermostat de-energizes a circuit and diverts all the heat to the heat pump circuit. The conventional water heater would then heat the water.

Phoenix Air Conditioning is the only heat pump manufacturer I know of that includes a desuperheater as a standard fitting and not as an extra. Wherever heat pumps or any application of the heat

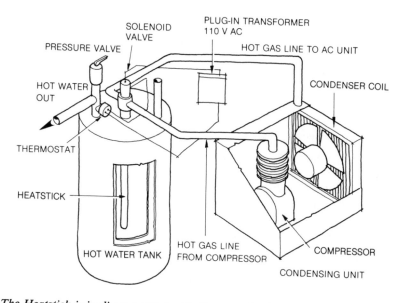

The Heatstick is in direct contact with the water in the hot water tank. Note the flow of hot gas from compressor to Heatstick.

pump principle is put to use, there is always the potential for the use of desuperheaters. They should be included in any new installation, and existing installations should be retrofitted if possible.

Most desuperheaters take the water to be preheated to the heat pump where there is a small refrigerant-to-water heat exchanger. The "Heatstick" made by Enro Manufacturing Co. Inc., Riviera Beach, FL 33404, takes the hot gas straight from the compressor to a heat exchange stick installed in the hot water tank in direct contact with the water. The Heatstick and water tank combine to form the desuperheater.

The manufacturer claims that the Heatstick will provide approximately ten gallons of 160° F. water per hour for each ton of heat pump or air-conditioning capacity. The average domestic heat pump is between two and four tons.

The cost of the Heatstick, distributed by Sun Harvesters, Inc., 416 N.E. Osceola Street, Ocala, FL 32670 ranges from $349 to $369.

Direct-Fired Heat Pumps

In the experimental stage are heat pumps that use conventional sources of heat, but have their own sources of power to operate the compressor of the heat pump.

A solar-powered heat pump is being developed by Donald Frieling and Associates at Battelle Institute in Columbus, Ohio. The system uses an array of solar collectors to heat vapor and thus power turbine vanes which in turn power the heat pump rotary compressor. The vapor spins the vanes at between 1,800 and 3,600 revolutions per minute.

The Gas Research Institute and the Department of Energy (DOE) are sponsoring the development of a gas-fired heat pump by General Electric. It will use a gas-fired sterling cycle engine to power the refrigerant cycle. The goal is for a COP of 1.9 in the heating mode and 1.0 in cooling.

The Gas Research Institute and DOE are also sponsoring the

development of a gas-fired absorption cycle heat pump. This is being developed by Allied Chemicals and Phillips Engineering of St. Joseph, Michigan. The Allied-Phillips unit is expected to operate at a COP of 1.3 in the heating mode and 0.5 in the cooling mode when it is developed.

The absorption cycle is very different from the compression cycle heat pump. In stead of a piston, the heart of the absorption unit is a generator. Heat from a burner, powered by oil, coal, or solar energy, activates the generator and starts the heating and cooling process. The generator is basically a boiler. The refrigerant, usually ammonia as opposed to Freon in compression heat pumps, needs heat to boil into a vapor. Absorption heat pumps consist of a generator, evaporator, condenser, and absorber.

Absorption cycle air conditioners for cooling only are made by Arkla Industries, Inc., Evansville, Indiana 47704. The firm's sales office is at 400 E. Capital, Little Rock, AK 72203.

Heat Pump Components

Heat pumps can be broken down into a few basic parts, each with its own job to do in heating or cooling the home.

These basic parts are:

1. *Two heat exchange coils.* One is the condenser, to give off heat in the home. The other is the evaporator, to pick up heat from the outside. In the cooling phase, the roles of these two coils are reversed.
2. *A compressor.* This compresses the refrigerant to increase its temperature.
3. *An expansion valve.* This does the reverse of the compressor. It permits the refrigerant to expand and thus drop in temperature.
4. *Fans or pumps.* There is one fan or pump with each heat exchange coil, to speed the process of accumulating or giving off heat.
5. *An accumulator.* It holds a supply of liquid refrigerant.
6. *Four-way valve.* It permits the heat pump to be switched from the heating to the cooling cycle, and back again.
7. *Resistance heater.* This auxiliary heater is housed in the condenser unit, and is used to help heat the home when the heat pump cannot provide enough heat. It also switches on during the defrost cycle, to remove ice from the coil outdoors.
8. *Controls, casing, and mounting.*

9. *Refrigerant.* This is a liquid with an extremely low boiling point, so that it changes from liquid to vapor as it picks up heat and is compressed, and returns to liquid when it gives up that heat and is decompressed.

Heat exchange coils

Heat exchange coils on air-source heat pumps are usually constructed of a galvanized frame and aluminum fins, with copper tubing. The fins are vertically spaced, ten to an inch for efficient air flow. Internally there are rows of copper refrigerant tubing running serpentine-style horizontally from top to bottom. Each row has a staggered placement for maximum air flow exposure to refrigerant tubes, which eliminates the shadow effect of air flow to each row of refrigerant tubes. The exact arrangement of the fins and tubes differs from model to model.

Some coils feature a venturi distributor to insure equal refrigerant flow from the expansion valve to all of the refrigerant tubes.

Refrigerant-water exchangers

Exchangers working between refrigerant and *water* are completely different. Most are tube-in-tube counter flow heat exchangers. On the Phoenix and Vanguard models, polybutyline water casing is used for the outer tube.

The casing expands and contracts during heat transfer cycles, thereby discouraging formation of scale on the casing. Water line connectors are soldered with high grade lead solder.

The inner tube carrying the refrigerant is cupronickel, 90 percent copper and 10 percent nickel alloy, and highly resistant to corrosion. Copper "Y" fittings are silver brazed to the cupronickel tube. Phoenix manufactures twelve feet of heat exchanger for each ton of refrigerant capacity, so on the three-ton PET 36 unit there is a thirty-six-foot, water-to-refrigerant coil. Because it expands and contracts with different water temperatures, cupronickel resists the build-up of mineral deposits from the water. As such it is superior to copper.

Mineral deposits are only a problem in areas where the water has a high mineral content; the extreme can be seen in Hot Springs, Arkansas, where mineral deposits can clog two- or three-inch diameter pipe in a short time. Consideration should be given to cleaning out deposits where necessary.

W.R. Heat Pumps, Ltd., an English manufacturer of heat pumps for swimming pools, uses a tube-in-shell, water-to-refrigerant heat exchanger. A coiled tube containing the refrigerant is inserted in a glass fiber water container. The two fit like hand in glove. This affords access to the water shell for cleaning.

Refrigeration Research, Inc. (Michigan) manufactures a water-to-refrigerant evaporator. The water coil is of integral finned copper tubing and the refrigerant expands into the steel shell which forms the outer casing.

Compressors

The compressor is the heart of the heat pump. It maintains the beat of the machine and keeps the refrigerant circulating within the cycle.

Compressors have to contend with a wide difference in temperatures between condenser and evaporator. In winter, outside air temperatures can fall well below 17° F. while outdoor summer temperature may be well above 100° F. The pressure and temperature of the circulating refrigerant vary enormously under such climatic extremes. To design a compressor to cope effectively with these changes is no small engineering task.

Back in the early fifties when heat pumps were introduced, they had a reputation for poor reliability. Compressors failed, valves leaked, and wires frayed. But corrective measures have been taken, such as the heavy-duty heat pump compressor, the addition of suction line accumulators, and improved crankcase heaters.

Types of compressors
The two types of compressors in use today are the piston compressor and the rotary compressor.

The *piston* compressor, also known as a "reciprocating compressor," draws the suction gas in at a low pressure and low temperature from the evaporator. An electric motor drives the piston to compress the gas. At the highest point of compression a discharge valve releases the gas to the condenser. The pressure, and hence the temperature, of the gas has been radically increased as a result of the action of compression. The high temperature gas is then moved into the condenser.

The *rotary* compressor performs the same function but in a more innovative way. It is about half the size and the weight of a piston compressor. The rotary compressor is composed of a roller in a casing which revolves somewhat like a loose ball bearing. It traps gas at a low pressure on the initial stage of its roll, and then as it continues, the pressure of the gas is increased until finally the gas is forced out of the discharge side of the compressor at a much higher temperature and pressure. Although manufacturers claim many benefits for the rotary compressor, I am not aware of any difference in efficiency between the two; rotary compressors do have the potential for higher efficiency because there is no need to convert rotary action to linear action, as has to be done with piston compressors.

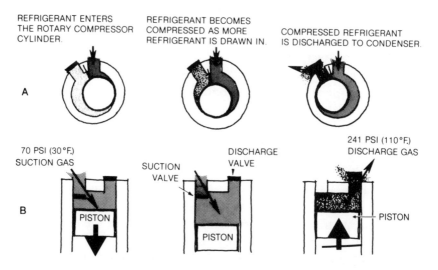

A shows compression cycle of rotary compressor. B shows piston action, drawing in gas, compressing it as piston moves up, then discharging it.

The only heat pump maker who claims "almost noiseless" operation is Solar Oriented Environmental Systems. Most compressors make noise so they should be well insulated and sited away from rooms that demand quiet (such as bedrooms). Water pumps, used to circulate water to and from water heat pumps (and swimming pools), can be just as noisy, and should also be sited carefully.

Expansion valves

Thermostatic expansion valves and capillary tubes are both designed to regulate the flow of liquid refrigerant from the condenser to the evaporator in exact proportion to the rate of evaporation of the refrigerant in the evaporator.

The amount of gas leaving the evaporator can be controlled since the valve can respond to:

1. The temperature of the gas leaving the evaporator, which can be sensed by a refrigerant-charged remote bulb
2. The pressure of the refrigerant leaving the evaporator, which can be sensed by a refrigerant-charged equilizer tube

The valve accurately responds to changing refrigerant temperature and pressure, thus allowing a more effective use of the heat exchanger as an evaporator.

Before passing through the expansion valve, the refrigerant is under high pressure on the condenser side. As the refrigerant passes through the expansion valve, the pressure and hence the temperature is reduced, thus enabling a good heat exchange to take place in the evaporator.

Four-way reversing valve

The electrically actuated reversing valve changes the heat pump cycle from heating to cooling by reversing the flow of refrigerant and thus changing the function of the two coils, with the outdoor coil changing from evaporator to condenser, and the indoor coil

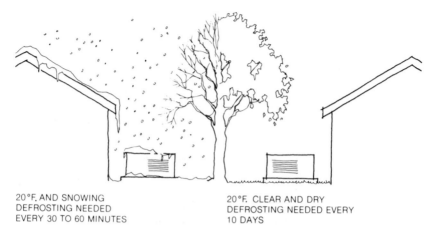

20°F. AND SNOWING
DEFROSTING NEEDED
EVERY 30 TO 60 MINUTES

20°F. CLEAR AND DRY
DEFROSTING NEEDED EVERY
10 DAYS

Weather conditions strongly influence how often heat pump must switch to defrost cycle and remove ice from coil in unit outside the home.

changing from condenser to evaporator. This valve also switches the heat pump to the defrost cycle when the outdoor coil becomes iced over.

Controls

Additional electrical controls are used to operate the following:

- Thermostat
- Defrost cycle
- Resistance heater
- Compressor motor and fan pump
- Reversing valve relay
- Overheat protection

In addition there are separate controls for desuperheaters and for add-on heat pumps.

Of the items listed above, the one that uses the most energy is the automatic resistance heater, used both for defrosting and for heating when the heat pump cannot meet the heat demand of the house. Later on in this book we will discuss ways to reduce or eliminate the need for this item.

A new heat pump *modulator* control system was introduced in 1980 by York to go with its Champion range. The manufacturer claims the modulator will reduce heat pump energy use by 17 percent by varying compressor and indoor fan speed to suit heating and cooling needs. In most models an off/on cycle is used during operation. The York control system modulates the fixed utility-operated sixty-Hertz current to lower frequencies, thereby reducing motor speeds on the compressor and fan to conserve energy when heating or cooling loads are below peak design capacity.

In addition the control package includes a panel with light-emitting diodes which light up when any part of the system malfunctions, such as when the refrigerant charge gets too low. This facilitates servicing. The same keyboard replaces the conventional room thermostat; the owner need only key in temperature and times for automatic control.

One dealer with whom I discussed controls suggested that the growing complexity of controls and circuitry, especially with split-system pumps, could be reduced by using preheating systems together with separate back-up heaters burning gas, coal, or wood. He believed such systems would be cheaper and more efficient.

Defrosting

Rain, snow, and high humidity saturate the air with moisture. In taking heat out of the air, the evaporator causes condensation which quickly turns to ice even when surrounding air temperatures are above 32° F.

Defrosting systems have been improved in the past twenty years, although they still depend on reversing the refrigerant flow to melt the ice. This of course takes heat out of the house to remove the ice. Estimates vary on how much energy this wastes, but average between 5 and 6 percent of the annual energy required by the compressor. This does not include the additional energy used to supply resistance heating which is frequently turned on automatically during defrost or when indoor thermostats show the need for supplementary heat. This can be as high as 8 percent of total energy used.

Most heat pumps defrost on an improved time-temperature basis. That is whenever the temperature of the refrigerant in the outdoor coil falls below a pre-set level, the unit defrosts, provided a certain time has elapsed since the last defrost cycle. The cycle ends when a rise in temperature of the refrigerant indicates the ice has melted. In addition, the unit is set so that the length of time of the defrost cycle is limited.

Many manufacturers, including York, have moved to an improved type of control. The York system senses whether the evaporator really is frosted over, taking into consideration not only temperature but air humidity. If relative humidity is below 60 percent, coil freezing is rarely a problem.

In addition to low temperatures and high humidity, a third factor contributes to the problem of frost and ice: the wind-chill factor. A wind of ten miles per hour at 17° F. causes far less ice to form on the evaporator coil than a wind of twenty-five miles per hour at the same temperature. The wind does not change the temperature of the air but it has a chilling effect on the evaporator coil.

Wind Chill Index

WIND SPEED	AIR TEMPERATURE						
	40° F.	30° F.	20° F.	10° F.	0° F.	−10° F.	−20° F.
	EQUIVALENT TEMPERATURE IN WIND						
5 mph	37° F.	27° F.	16° F.	7° F.	6° F.	−15° F.	−26° F.
10 mph	28° F.	16° F.	2° F.	− 9° F.	−22° F.	−31° F.	−45° F.
15 mph	22° F.	11° F.	−6° F.	−18° F.	−33° F.	−45° F.	−60° F.
20 mph	18° F.	3° F.	−9° F.	−24° F.	−40° F.	−52° F.	−68° F.
30 mph	13° F.	−2° F.	−18° F.	−33° F.	−49° F.	−63° F.	−78° F.
40 mph	10° F.	−6° F.	−22° F.	−36° F.	−54° F.	−69° F.	−87° F.

This effect can be almost completely negated and the air-source temperature greatly increased by using passive solar shelters at the evaporator coil. In areas with good winter sunshine, these would reduce the need for defrosting. The need for nighttime defrosting would be further decreased by using a heat sink of rocks, water, or chemicals into which heat would be pumped during the daytime. This would not only reduce the need for defrosting, but would increase the efficiency of the heat pump.

Accumulator

Suction accumulators are small storage cylinders which contain liquid refrigerant from the evaporator and prevent it from flowing into the compressor suction line before evaporating. This improves reliability by removing strain on the compressor.

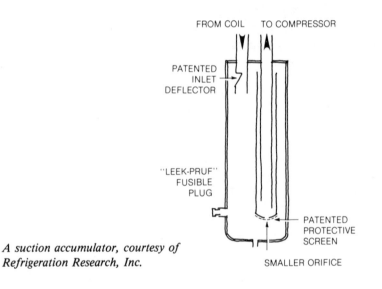

A suction accumulator, courtesy of Refrigeration Research, Inc.

Other components

Other components found in some units include the following:

1. *Filter-dryer.* This filter in the refrigerant line removes small foreign particles, and its dryer removes any moisture.

2. *Crankcase heater.* This boils off liquid refrigerant in the compressor and maintains a reasonably constant compressor temperature.
3. *Warning light.* In an air-to-air heat pump with resistance heating for back-up, a warning light on the indoor thermostat tells you when you are using the more expensive back-up heat.

Installation

Installation of a heat pump is not a job for the do-it-yourselfer. Most dealers will quote a price for a unit installed. This quote will include the usual installation costs, plus any extras, such as beefing up the home electrical system or improving the hot-air ducting system.

Outdoor coil

The difference between packaged and split systems is that with a split system, the outside coil can be placed anywhere, while the packaged system must be placed in a wall or roof.

Placement of the outdoor coil is linked with the efficiency of operation of the heat pump. For best results, the outdoor coil should be exposed to the winter sun for heat gain, shielded from the winter wind, and raised six inches above the maximum snow level.

For best heat dissipation during the cooling cycle, the outdoor coil should be shielded from the high and hot summer sun, and if possible placed in a cool, dark area.

Deciduous bushes and trees planted around the outdoor unit will prevent unwanted heat gain from the summer sun and create a relatively dark, cool area. In winter, after the leaves have fallen, the coil will be exposed to the heat of the low winter sun, which

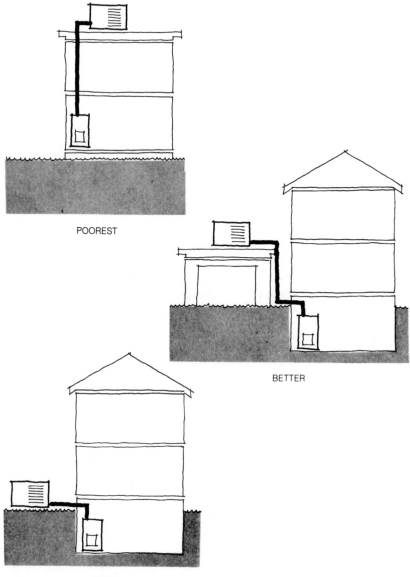

POOREST

BETTER

BEST

A roof arrangement exposes the coil to the summer sun and the winter wind. A unit mounted atop a garage is better. The south-facing location, shrouded in deciduous trees and bushes is best for both winter and summer.

can give a useful heat gain and reduce the need for defrost cycles.

The ideal arrangement, while expensive, might be to have two outdoor coils. The one for winter heating would have a southern exposure and be enclosed in a small greenhouse shell to protect it from the wind and increase the heat gain. The other, used to dissipate heat during the summer, would be shaded by trees or shrubs.

Water-source pumps do not present the same problem. The only difficulty is that the cooled well or lake water, pumped to a packaged or split system in the house, might freeze on the way back to its source or to a discharge well.

Inside coil

The indoor section of the heat pump can be placed anywhere in the house from the attic to the basement. It seems to make little difference where it is put. However, there can be heat loss if the indoor and outdoor coils are placed too far apart.

Ducting

The air distribution system for air-to-air water-to-air heat pumps must be larger than those for standard gas or oil systems because the heat pump provides hot air at temperatures ranging from 85° to 110° F., compared to 110° to 130° F. for oil and gas.

This means wider ducts and extra registers are needed. A rule of thumb is to use up to 50 percent more supply registers, and air ducts sized to carry 400 cubic feet a minute for each 1,200 Btuh of cooling capacity. Various types of forced-air heating systems are available, but there is no real difference in heating efficiency as they all use convection. What is important is that whoever designs the system, whether it be for a single residence or a tract, ensures that the supply and return ducts are properly proportioned for heat pump use. The major cause of unsatisfactory heat pump performance is inadequate or improperly routed air systems.

Hydronic or wet radiators need to be sized similarly. That is, the face surface area of the radiators needs to be 1.5 times that used with a gas- or oil-fired system. The exact dimensions of the hydronic system will vary with the design temperatures of the installation. In new construction, water-based heating systems can be used for underfloor heating.

PART II

HEAT
PUMP
SYSTEMS

*"A heat pump can use the
earth as both a heat source
and a heat dump."*

Air-Source Heat Pump Systems

Recently the sales of heat pumps have soared; half a million of them were sold in 1979 and most of these were air-to-air units. The reasons for this sudden increase are as follows:

1. Heat pumps have recently gained a reputation for being as reliable as other heating and cooling systems.
2. An air-to-air heat pump uses less energy than other heating systems.
3. When compared to electric resistance or oil heating, the cost savings is substantial.
4. Although the initial cost of the heat pump is slightly higher than for a conventional system, the payback period is usually less than three or four years.

The COP (coefficient of performance) with an outside air temperature of 50° F. is around three for the above-average models. As the outdoor air temperature falls below freezing (32° F.), the COP drops to about two, which is still a good economic proposition. Below 32° F. it is more difficult for the heat pump to extract heat from the freezing air. Frost collects on the evaporator, and a defrost system has to be turned on, which means heat-pump-heated air cycles from the house to the outside coil.

The economics of air-to-air heat pumps still look good down to between 10° F. and 20° F.; below this point the electric resistance heating system is automatically turned on. Although heat pumps can continue to operate at very low temperatures and with a COP of one (the equivalent of resistance heating), doing so is inefficient and puts a severe strain on the compressor. For this reason many homeowners have thought that heat pumps were made only for southern climates; however, they can be adapted for use in cold areas of the country.

The best solution is to encorporate passive solar gain in your heat pump installation. Passive solar principles, applied to northern installations, will reduce strain on the compressor and boost heat pump efficiency.

Another, more common solution is to plan an alternate heat source (wood, coal, oil, or gas) to take over whenever the heat pump is having to rely heavily on the electric resistance back-up. This allows the alternate fuel to provide heat during the coldest weather when it can be burned most efficiently.

Air-to-Air:
A Vermont Installation

Although not as numerous as they are in the South, heat pumps in the North have been increasing in popularity for the past eighteen years. Many northern homeowners, however, still fallaciously assume that heat pumps are unsuited to climates with freezing winters and almost-warm summers. To see how well they do work, I thought I would check with a northern dealer.

Dee and Mike Kilburn have a family electrical, refrigeration, air-conditioning, and heating business in Manchester, Vermont. They run the business out of the family home, a big, rambling three-story Victorian house with a large wrap-around front porch and a four-car garage. They got into the heat pump business for very practical reasons.

The Kilburns have a work force of seven, and the volume of their work is immediately affected whenever there is a slump in the construction industry. During the 1973 construction slowdown, Dee was forced, for the first time, to lay off some of his

Mike Kilburn and his son inspect the two exterior units that combine to both heat and cool this three-story house in Manchester, Vermont.

employees. He didn't like laying off workers, particularly since they might not be available to him when he was ready to rehire; he knew he had to diversify. He wanted an item with a future, one that he could install, service, and maintain, and which would help provide year-round employment for his employees. He decided to expand the commercial air-conditioning end of the business since its peak time generally occurred during the times when his electrical business slowed. Heat pumps, air conditioners that reverse their function, became a natural extension of this decision.

Personal reason

The second reason was personal. In 1978 the Kilburns burned 2,200 gallons of oil to heat the 2,600 square feet of their house. Even that was a decrease. Kilburn chuckles when he recalls, "We once had it up to 3,000 gallons."

That was before they started taking measures to decrease fuel consumption. Over a seven-year period they added blown cellulose to the attic, replaced the boiler, and nailed one-inch foam insulation to the exterior walls of the first floor before sheathing them with vinyl siding. Next Kilburn started thinking about wood heat, but he knew it was going to be expensive. Not only was there the cost of a wood-fired boiler, but there would be the added expense for three stories of masonry, or a prefabricated, insulated chimney.

Kilburn also considered installing a heat pump, which from his commercial installations he knew worked well. The few units that he had installed in vacation homes were also satisfactory. However, he didn't have first-hand information on how well they would work in a full-time residence, nor how cost-effective they were, given the initial price tag. In short, he asked himself, is a heat pump, when looked at as strictly a heating system, cost-effective and practical in a northern climate?

Feasibility

Mike, Dee's son, spent a lot of time gathering degree-hour figures for Vermont because he knew that heat pumps are not as efficient when it is extremely cold. As the temperature drops, there is less heat in the outside air for the heat pump, and at the same time the house needs more heat.

Much to his amazement, Mike discovered that about three-quarters of the degree-hours in Vermont occur above 15° F., making northern climates theoretically warmer than they feel. He also discovered that, although there is still little reliable comparative data on heat pump compressor failures, it appears that temperature has little effect on the function of the components.

Next Dee set Mike to figuring heat loss/heat gain calculations. Dee knew that if you figured the Btu gain of a house, you could determine the size of the heat pump required, and from there calculate the operating costs. (You can do heat loss calculations yourself, using *The Complete Energy-Saving Home Improvement Guide,* available from local state energy offices. Audit teams from the State Extension Service will also do them for you, or you can ask your heat pump dealer, who provides this information as a service.) Mike's results looked good — but nowhere near as good as they turned out to be.

Ductwork

Once Dee decided to go with a heat pump, ductwork was an immediate problem. There were no ducts in the house. They would have to be added, and this meant taking out some walls, a major task.

So Dee decided to take another route. He elected to install two heat pumps, one in the cellar to be ducted into the first-floor registers, and a second in the attic for the second-floor rooms.

The ductwork he used (and what he uses for all his installations) is a rigid insulated fiberglass board which comes in large, flat sheets. There are special tools to rabbet the edges for interlocking seams. Then the board is scored to the desired dimensions, and folded. The rectangular-shaped duct is stapled and taped together. Once all the duct-shaped lengths are made, they are assembled on site. The end rabbets interlock; the joints are stapled and taped.

This ductwork is light and, because it is made of an insulating material, heat loss is small. The Kilburns had no intention of branching out *quite* this far, but they found that by making their own ductwork, they could insure that the ducts were built to the proper dimensions and installed correctly — important, as you'll learn, to keep noise levels low.

Kilburn estimates his system with two heat pumps cost $6,500 installed. It sounds like a lot, but to Kilburn it's the best investment he's ever made. In one system he gets lower fuel costs, plus heating, cooling, air filtration, dehumidification, and the added possibility of air humidification.

Electric Rates

From Kilburn's figures-to-date, the payback period is only a year or two away. After thirteen months, he has used 18,310 kwh of electricity to run the pumps at a cost of $457.75. (He put in a separate electric meter to doublecheck the manufacturer's figures.)

Electric rates are important to the cost-effectiveness of a heat pump. Manchester, Vermont, has peak and off-peak (or time of day) rates. Off-peak rates are $0.025/kwh and peak rates (between 8 and 11 a.m. and 4 and 9 p.m.) are $0.14/kwh for the four winter months, or two-thirds of the heating season.

Heat pumps are particularly compatible with such a rate structure. Because daytime temperatures usually are higher, the heat pump will operate more efficiently and use less electricity than at night. In addition, you may be able to avoid peak rates entirely. "We keep the downstairs thermostat set at 70° F.," Mike said. "If there's no sun, no wind, and outdoor temperatures are above 30° F., we can coast through the period of peak rates by setting the oil-fired boiler (the back-up system) at 60° F. If it's colder than 30° F. outside, or is windy (the exterior walls of the second floor still aren't insulated), then the oil comes on." But with a total oil bill for 1979–80 of $125 (150 gallons of oil), the oil doesn't come on much.

Their heat pumps have a COP at 47° F. of 2.7–2.8, with a computed balance point of 27° F. At that temperature, the output of the heat pump is equal to the heat requirement of the house. Even at 17° F., the COP is 1.9–2.

Now that the Kilburns have some firm data from their own installation, they have begun a more aggressive approach to residential as well as commercial installations. For commercial installations, they get balance points of around 0° F., which are possible because commercial establishments generate some of their own heat from pedestrian traffic, appliances, and lighting.

Is a heat pump for you?

The Kilburns agree that heat pumps should be considered for *any* new construction, or whenever an old heating system nears replacement age, or when central air conditioning is desired. Homes that are the most difficult to adapt for heat pumps have non-ducted heating systems, such as hot water or steam. These require extensive and expensive restructuring of the house. Passive solar homes set on a slab floor may be even worse because usually there is no space for ducts. A home with an existing gas, oil, or electric forced hot-air system will be much easier. The air distribution system is already in place; the heat pump need only be fitted into the supply air ducts of the forced-air system. There is, however, one large proviso: ductwork *must* be properly sized.

Unfortunately many homes have inadequate ductwork, especially if the ducts were sized without any expectation of installing central air conditioning. In addition, installers have been known to cut corners on hot-air systems by reducing the number and/or size of the ducts. Because heat pumps generate low temperature heat, accurate sizing and placement of ducts and registers are vital. This may mean adding supplemental ductwork or increasing duct size.

If the equipment for the heat pump is oversized, the house, when the heat pump is running on the air-conditioning cycle, will feel cold and clammy. If there aren't enough ducts or if they are poorly located, air flow may be insufficient, or you may find yourself constantly sitting in cold drafts.

The Kilburns had to put in a second panel box for the heat pump electrical hookup, but that, says Dee, is fairly usual in residential installations. The heat pump needs a 200-amp service; most houses have only 100 amps.

Sizing a heat pump

Heat pumps are sized on the basis of prevailing climatology; heat loss figures; and on the cooling, not heating, requirements of the home. This is fairly confusing for northerners, but remember that heat pumps were originally designed as cooling units.

Sizes are given in tons; a one-ton heat pump will remove 12,000 Btu an hour. This translates, in the heating mode, to 12,000 Btu per hour of heat input, plus whatever heat is given off by the unit. A two-ton pump will remove 24,000 Btu per hour, and a three-ton, 36,000 Btu per hour. Most homes will require a two- to five-ton unit.

A heat pump, sized for the air-conditioning load, generally achieves a balance point around 27° F. and even though the heat pump efficiency declines as the outdoor temperature declines, its seasonal COP or efficiency with supplemental heat will still be in the 1.7–1.8 range. In other words, with a COP of 1.7 and a time-of-day rate of $0.03/kwh for the overall heating season, the equivalent oil cost at 60 percent efficiency would be $0.43 per gallon.

In most applications heat pumps are designed so that when there isn't enough heat in the outdoor air, the unit automatically calls in supplemental resistance heat, which warms the circulating air. The resistance heat is actually the second or third stage of a multi-stage thermostat. This means that the heat pump continues to run while calling in an occasional boost of heat from the resistance heater. The length of time the resistance heat stays on depends upon how far below the balance point the outdoor temperature has dropped. In other words, the colder it is, the longer the resistance heat will stay on. It is important to remember that the heat pump does not shut down when the supplemental heat comes on. The reason is that although the heat pump may not produce enough heat to heat the home, it still provides heat more efficiently than a conventional resistance system.

For example, if it were 17° F. outside and the heat pump system had a balance point of 27° F., the heat pump would require some supplemental heat since the outdoor temperature is 10° below the balance point. However, the heat pump most probably would still have a COP of near two, which means it could produce twice as much heat as resistance heat using the same amount of electricity.

A heat pump operating on time-of-day rates usually has a back-up heating system in northern climates, to provide an alternate source of heat during peak hours. This can be as simple as a wood stove or as complex as a whole house heating system. The alternate system gives the homeowner the added advantage of being able to select the most cost-effective fuel for a given time. The homeowner will be utilizing the heat pump for cooling in the summer and for its high heating efficiency during mild winter temperatures; during the very coldest weather the second system will kick on when the full capacity of its furnace can be used to its best advantage. Quite conveniently, a gas, oil, wood, or coal heating unit is most efficient at very cold temperatures because the furnace remains "on" most of the time. Efficiency is severely decreased when the unit continuously cycles "off" and "on."

Plans wood back-up

Mike, who is working on figures for a house he is building for himself, intends to use a heat pump in conjunction with wood back-up for peak hours and below 0° F. heating.

"Wood," he points out, "is a perfect back-up for a heat pump. Woodburners find that they're usually too hot in spring and fall. So they dampen down the fire; this not only means incomplete combustion, decreased heating efficiency, but also tremendous creosote buildup. Wood stoves or furnaces work best in the middle of the heating season, when temperatures are 10° F. and below and the stoves and furnaces are burning hot with the draft wide open. That's when the heat pump is most inefficient. It's a perfect combination."

Location

The Kilburns have learned a number of things from their installation. Like many people, they chose a split system. Packaged units usually have a lower COP, and they are noisier, which in a residential dwelling can be extremely annoying.

"We may have become accustomed to the noise of vacuum cleaners, disposals, and mixers, but the noise of a forced air heating system — and that's what a heat pump is — is especially nerve-racking because heating systems run day and night."

The split system isolates the noise in the outdoor unit. This part of the system takes in air on four sides and discharges it at the top. Result: both discharge air and sound are directed upwards rather than towards the house. Such a design also protects nearby plantings from the damaging rush of hot, dry air during the cooling phase.

The Kilburns located the two outdoor coils on the north side of the front porch. "It's for advertising," Mike says with a smile, knowing full well that although the units may be close to the street, they're hidden by shrubbery. But the Kilburns think that even in this location, the heat pumps could create too much noise if the adjoining room were a bedroom.

Solution

Mike tersely explains the noise problem and what to do about it. There are two possible sources of noise. One is the rush of air inside the ductwork. Close attention to ductwork design and terminal fittings will eliminate this problem. The other is the outdoor unit, moving a large amount of air to achieve a decent operating efficiency. The noise is from the fan moving this air. Proper placement of the unit will minimize the problem. Don't place it near a bedroom; it won't be troublesome outside a laundry room or kitchen, he promises.

Drafts

One complaint the Kilburns have heard from heat pump aficionados is the cold draft from registers. Mike explains that we are used to forced-air systems that blow 140°–160° F. air. Drafts of this temperature are comfortable, but the forced air from a heat pump may be only 80° F. Since this is well below body temperature, it will feel cold if you are sitting near a register. Careful design and placement of ducts should prevent this.

Defrost

During the heating cycle and under normal conditions of high humidity or freezing rain (which occurs at 30°–40° F. outdoor temperatures), there may be a frost buildup on the outdoor coil of air-to-air units. As the ambient air temperature falls, frost accumulates on the evaporator coil exterior due to freezing of the moisture which condenses out of the air as it is cooled. A certain amount of ice formation is tolerable, but after a period of time, the buildup will be such that not only does the ice form a thermal barrier between air and refrigerant but also it will completely obstruct the passage of air, and the evaporator will cease to function as a useful heat source. If frost is permitted to accumulate, the effectiveness of the evaporator in removing heat becomes severely reduced. It is necessary to remove the ice rapidly at that time.

The unit senses this buildup of frost, and automatically reverses itself and goes into a defrost operation. Normally this lasts for no more than a few minutes. In northern climates the heat pump will defrost more frequently as the temperature falls. You may know when the defrost cycle is occurring; supply air will briefly feel a little cool.

Choosing a dealer

Choose a reputable dealer, says Kilburn. You need an installer with technical knowledge and know-how. If the heat pump is not installed properly, or is not kept at peak operating efficiency, you will, at best, suffer lower than optimal COP, at worst you will have a system failure. Dealers can be found listed in the Yellow Pages under the Heat Pump section.

A dealer should be chosen who has proven competence, training, and accreditation by the manufacturer, and who will be available to service the heat pump twenty-four hours a day. Also be sure that before the pump is installed, the dealer or contractor does a complete heat loss/heat gain study of your house in order to size the equipment properly.

Kilburn also stresses the importance of the service agreement. The heat pumps handled by the Kilburns have an automatic five-year warranty on the compressor, and with all General Electric models, it is possible to buy a five-year service agreement for parts and labor.

Maintenance of the units is essential and should be done by trained personnel. Filters on the forced-air systems will have to be cleaned or changed at regular intervals, and the machinery should be oiled and cleaned annually.

After we walked around to see the various heat pump components, Mike took me into the office where they have their computer which can do so much for the person wondering whether to purchase a heat pump.

Mike had promised to show me how heat loss and payback figures are computed. He dialed a series of numbers on the telephone, which is connected with Cleveland, and from there the connection was automatically made to Amsterdam, Holland — G.E. computer headquarters. The phone mouth and ear pieces were fitted into rubberized receptors connected to the computer. We began feeding in information.

We were using a hypothetical (but average) New England house. To run payback figures, Mike needed to know the heating design load (70,000 Btu), the type of furnace (oil), the inside ambient temperature (70° F.), the efficiency of the oil burner (75 percent), the electric rate ($0.025), and the present cost of oil ($1.05). We would be considering an add-on unit. We also needed predicted

Cost and oil consumption dropped dramatically when heat pump was installed.

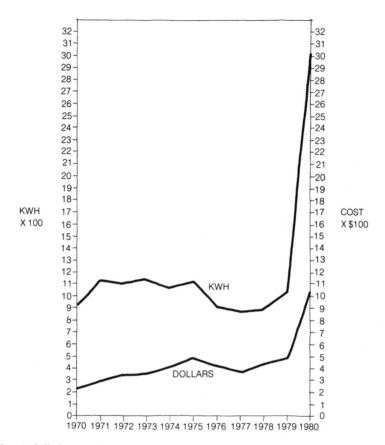

Electric bill shot up, but not as many dollars as oil bill dropped.

rate increases, so we added a 10-percent annual increase for electric rates and an 18-percent annual increase for oil over a ten-year period.

Payback period

The print-out had a payback period for an add-on costing $2,000 of 2.9 years; the payback period for the same house sized for heat pump only would be 3.14 years, which included the initial cost of the equipment and installation.

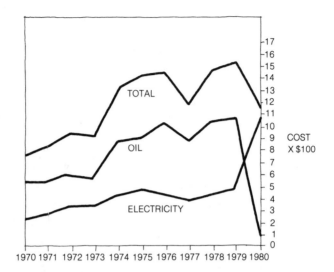

This graph tells the story: total costs down after heat pump goes to work.

The real shocker was the projected fuel figures for a ten-year period. This time we increased the ten-year rise in oil to 26 percent and electricity to 14 percent and came out with an annual oil bill of $12,000 in 1990. For that same year the electric bill for the heat pump was $1,800.

Even though the figures are hypothetical, the message is clear. At the present rate of price increases, other or supplemental heat sources are essential.

The same computer has other programs; it will print out heat loss figures as well as cost of and payback periods for different kinds of insulation.

Mike showed me the print-out for a home in which the Kilburns are planning to install a heat pump. The figures showed that if the homeowner insulated just around the windows and doors, more than $1,000 in oil costs would be saved each year. "If you spend your money weatherstripping and insulating," says Mike, "you can effect a permanent energy savings, and you may be able to make up the cost of insulation before the heat pump even begins operating. Sometimes insulation also means sizing the house for a smaller heat pump and smaller ducts. It pays to insulate first."

Mike and Dee Kilburn are excited by the potential applications of heat pump technology. Advanced heat pump systems are being developed that will incorporate a heat storage system: the hot water tank. The hot water will be heated by off-peak electricity (or solar collectors) and the stored heat used as a heat source for the heat pump in periods of extreme cold. This possibility may be an important factor in the widespread acceptability of the heat pump.

Also they eagerly await the two-speed pump that will be even more efficient. This heat pump will operate up to 90 percent of the time on low speed in cooling. The "low speed" setting will minimize off/on cycling, which in turn will contribute to a longer compressor life; compressors are under less stress at low speeds. This will also reduce start-up stresses because all starts will be at low speed and fewer starts will be required, thus saving energy.

During peak demand, the compressor will automatically shift to high speed. The electronic monitoring controls will be programmed to actuate time/temperature setbacks twice a day.

While the rest of us try to save energy by shivering in 65° F. or lower houses, Mike and Dee Kilburn are turning up the thermostat — and saving money. You can't tell them that Vermont winters are cold.

The Passive Principle

Northern homeowners can also increase the efficiency of their heat pumps by using the sun.

The weak winter sun may appear to hold little heat, but just let that sun shine into a conservatory, a solar green house, or a passive solar shell and suddenly weak sunlight has a powerful warmth. The same principle can be applied to heat pump usage by opening the evaporator to the winter sun while protecting it from the winds. This will cut winter heating costs in half or more, and reduce the strain on the compressor, thereby increasing the life of the heat pump.

To build a passive solar shell to house the outside coil can cost less than $100. What is required is single- or double-glazed glass; a

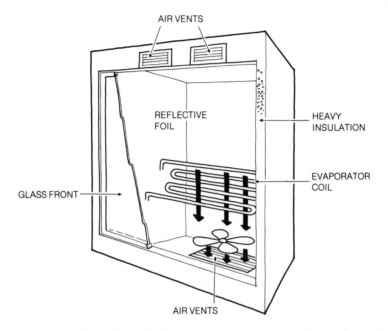

Passive solar shell is built with glass in front, so sun warms air entering top of shell. Air heats pipes containing refrigerant, then is pushed out the bottom of the shell by the fan.

wooden or aluminum frame to hold the glass; vents, such as louvered glass; and reflective foil or aluminum foil.

Make a concrete base for the heat pump on the south side of the house where there is maximum solar exposure in winter. Insulate beneath the base, and cover the sides of the shell with glazing.

The purpose of the glazing is threefold:

1. To preheat incoming air
2. To facilitate a direct solar heat gain on the heat pump itself
3. To eliminate the wind chill factor by creating a "tea cozy" effect. The wind chill factor can substantially reduce the efficiency of a heat pump. By protecting the outdoor coil with a passive solar shell, the wind chill factor is reduced, if not eliminated.

The solar shell will preheat the incoming air during daylight hours; the degree of preheating will be slight if the shell is tight-fitting around the heat pump and if the air flow is directly to the coil, or it can be a substantial heat gain if the air flow is directed through a series of heat-absorbing tubes in or below the passive shell.

The greatest heat gain is from direct solar radiation heating the evaporator coil. Additional radiant solar energy is available by reflection if foil is put on the northern side of the passive shell.

The radiant solar energy raises the temperature of the evaporator section by means of the *greenhouse effect*; short wave radiation from the sun passes through the glass, but is converted to long wave radiation that cannot get out and so remains as heat.

The passive solar shell need only be a few inches larger than the outside coil with fan, but would contribute more to air preheating if it were bigger. A flat plate solar collector, with an insulated metal back, metal or wooden sides, and glazed, would serve well as a solar preheater for air entering the passive heat pump shell. No fan would be required; however, the air would flow more easily if the panel were placed below the heat pump.

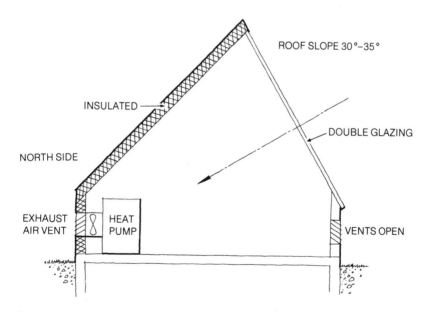

Solar greenhouse design for heat pump. For northern climates and for heat-only units. Note air enters vents on south side, exits on north side.

Perhaps the best place for the outside coil is in a lean-to solar greenhouse, or in the outer shell of an "envelope" solar house. Temperatures in lean-to greenhouses can reach 50° F. to 60° F. even in sub-zero outdoor conditions. Beause they are not used in winter, greenhouses can be the perfect source of solar preheated air for air-to-air heat pumps.

Thermal mass

The solar shell/heat pump arrangement should be located against a house wall to reduce heat loss from both the shell and the house. It also facilitates the direct connection of the refrigerant lines to the house, which eliminates heat loss from the lines.

For optimum solar gain and peak performance, the heat pump should be operated only during daylight hours. This necessitates thermal storage to absorb the heat during the heating cycle and to radiate it back out during the night. The house itself can be so well insulated as to transform everything inside the house into thermal storage units. The more the actual structure of the house itself is used as thermal mass — slab floors, brick walls, and masonry

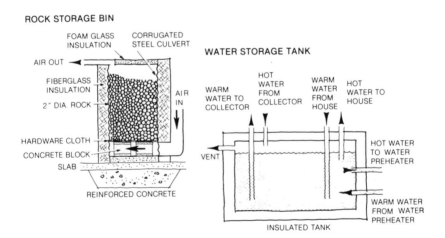

For storing heat, the water tank at right will hold more Btu than an insulated rock storage bin of the same size, and is better for hydronic systems. Rock storage is convenient for air heating systems.

fireplaces — the better the house is at retaining winter warmth and summer cooling.

Where additional night heating is required, the choices range from use of an alternate heat source to use of a twelve- or sixteen-hour thermal storage bin of rocks or water. Rock storage is more convenient for air heating systems and water for hydronic systems. Overnight heat stores cost no more than a little planning followed by hard work. Rock storage consists of clean, hard stones in an isolated container with ducts. Finding a basemement or part of the insulated structure of the house to put the storage in can be difficult.

A more detailed discussion of heat storage systems is given in Appendix C.

The passive potential

Passive solar shells used with heat pumps for agricultural, industrial, and domestic applications have tremendous potential. For example, if the air temperature is 35° F. and there is a fifteen-mph wind, the effect on the outdoor evaporator will be similar to that of 17° F. air. Some heat pumps have a COP of two at 17° F., but with the probable need for defrosting and additional resistance heating, the SPF drops to 1.5. Assume that a passive solar shell is built around the evaporator and that winter sunshine, diffuse, or direct, contributes a 10° F. heat gain, then the 35° F. air will be increased to 45° F. The result will be a COP of almost three, double what it would be without the shell. In addition, there will be no need for defrosting or auxiliary heating.

In northern climates having a lot of clear winter sunshine, the heat gain can be substantial, more than doubling the COP. Imagine what this could do for industrial or agricultural installations where available air temperatures at the evaporator may be as high as 65° F., and the process heat required is between 70° F. and 90° F. In such cases, properly designed heat pumps should be capable of reaching COP's as high as four and five.

In cases where heating and cooling are required, use two outdoor coils, one to serve as an evaporator and the other as a condenser. Connect both coils to the one indoor compressor and condenser. This allows the evaporator coil to be placed in a permanent solar shell in the optimum location for winter heat gain. The condenser coil may be placed on the north side of the building, in the coolest area possible, well shaded by bushes, shrubs, and trees. Such a location prevents sunshine from adding heat to the condenser, which is trying to throw off as much heat as possible; the cooler the air passing over the condenser, the higher the EER.

Solar greenhouse

Mention solar greenhouses to most people and the response is likely to be: "But aren't all greenhouses solar?" Not so. Ordinary greenhouses lose heat and can cost a small fortune to heat whereas solar greenhouses are in effect tightly constructed solar collectors which trap heat. The increased air temperature inside the solar greenhouse makes the use of air-source heat pumps in cold northern climates feasible.

The solar gain on a clear January day at a location on the 40th parallel is at the rate of 290 Btu/hr. per square foot at noon. At this rate enough energy would pass through a mere three square feet of a 12-by-16-foot solar greenhouse to raise the air

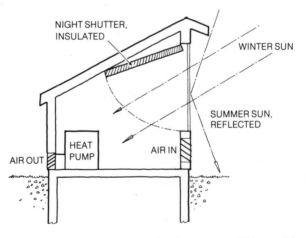

A solar greenhouse designed for use with a heat pump. This model can be used year-round for both heating and cooling.

temperature inside from 40° F. to 70° F. in just one hour. The total clear-day solar gain on a south-facing wall in January is 1,750 Btu per square foot, and the average net daily gain is 850 Btu per square foot after losses due to cloud cover are deducted. This means that a 10-by-20-foot solar greenhouse can gain the equivalent of 50 kw in heat.

Such a greenhouse can be the perfect "passive solar shell" for a heat pump. No design change is required other than the inclusion of adjustable vents to allow air in and a duct to take the air from the heat pump back outside. The air intake can be a section of the glass louvers frequently seen in kitchens and bathrooms. The air duct can be a section of standard heating and air-conditioning ductwork.

The design of solar greenhouses has now become standard. Two good books are available on the subject for those who want to pursue the idea:

The Solar Greenhouse Book, James McCullagh, ed., Rodale Press, 1978.
The Complete Greenhouse Book, Peter Clegg & Derry Watkins, Garden Way Publishing, 1978.

A Simple Subterranean System: Air-to-Air

I was on a drive to Eureka Springs in the Ozarks of North Arkansas when I saw a bright silver multi-blade windmill atop a small hill. The house was set in the hill, open only to the south and shaded from the summer sun by deciduous trees. I stopped and pulled up to the house where I met Sam Rainey and his wife, the owner-builders of this underground house.

It is a 1,488-square-foot retirement home and was completed in September 1979. Sam drafted the plans; he dug out the foundations, laid the floor, set the wall frames, and poured the roof. He also did all the wiring, plumbing, and cabinetry.

A 1½-ton split-system air-to-air Carrier heat pump supplies

This underground house has a windmill for water pumping and a heat pump for heating and cooling.

what little summer cooling is required. In addition, the heat pump is available as a back-up source of heat to the wood stove. Rainey made the ¼-inch steel stove himself. Last year he burned less than two cords of wood and did not have to use the heat pump at all.

Compressor on roof

The heat pump installation is the epitome of simplicity. The outdoor section, which includes the compressor, is mounted on the grass-covered roof. There are three feet of soil on top of the house. The outdoor unit is protected by a metal fence. Two refrigerant lines lead to the indoor heat exchanger in a utility room at the back of the house — the outdoor section is directly above the indoor unit. Air is ducted from the indoor unit throughout the house.

The heat pump is used to maintain comfort in summer with the assistance of room fans. The cooling EER is 8.8, the COP is rated at three at 47° F. air-source temperature and two at 17° F.

The total cost of the heat pump and installation was $1000 with Rainey doing most of the work and a heat pump dealer doing the final connections, checking it out, and charging the unit with refrigerant.

An alternative

The windmill will be used to pump water from a 300-foot well. The well supplies up to twenty-five gal./min., way in excess of domestic needs, but may be used for irrigation purposes. An alternative to the split-system air-to-air unit Rainey chose could have been a packaged water-to-air heat pump using the well water as a heat source/sink. This might be easier to install than a split system and certainly would be more economical to run.

Inside the underground house, this small section fits into a corner of the utility room and feeds the ducting system with either hot or cold air.

The outside unit of a heat pump in this underground house sits atop the house, on top of two blocks placed on a concrete circle. Refrigeration lines and electrical wire at right are run down through pipe into house.

The house is oriented towards the sun and while it is obviously well shaded from the summer sun, it is not specifically designed for passive solar gain. Its tight, well insulated construction, with wood heating and heat pump make it a remarkably energy efficient house.

SUMMARY

Location:	North Arkansas
Date of completion:	September 1979
System:	Split system air-to-air
Heat pump:	Carrier air-conditioning
SPF:	Unknown for this application
	Max. COP 3
Cost:	$1,000 installed

Attic Heat Space:
Air-to-Water

The heating system in the Keable house includes a heat pump with an evaporator coil in the attic (the attic acts as a passive solar trap) and a condenser coil in a basement thermal store. The heat pump and increased insulation combine to reduce house heating costs by 70 percent.

Background

This comprehensive exercise in energy conservation was carried out by Julian Keable, an architect, at his London home. Following the 1973 oil crisis, a competition was held with the Copper Development Association offering prizes for original ideas using renewable energy and, of course, copper. Keable and a refrigeration engineer, Christopher Dodson, proposed a scheme to optimize standard American heat pump equipment to heating-only operation. (The temperate London climate, similar to San Francisco, needs little or no cooling.) Their idea was to convert a standard roof into a simple solar collector, then pump the trapped heat to a thermal store in the basement.

The proposal received an award, and a grant from the Anglo-German Foundation was used to purchase the heat pump.

Insulation

The first step, paid for by Keable, was to insulate the house. He put six inches of fiberglass on the floor of the attic and covered the walls with three inches of fiberglass, topped with asbestos slates to keep the insulation material dry and retain the thermal mass of the walls. He left the front of the house as it was with its red brick Edwardian facade, and instead covered the inside walls with one inch of polyurethane.

Better Homes and Gardens

Insulation and a heat pump cut fuel bills in this London town house.

Next he draft-proofed the outside doors and added air locks at the front and back doors. Finally he blocked all fireplace flues and double-glazed the windows.

The heat pump system

No modifications to the structure of the house were required to fit the heat pump system, except for the minor addition of a cowl through which air is exhausted after the heat has been removed from it.

Outside air, the heat source, is drawn in under the eaves of the existing slate roof and channeled between the rafters by means of stapling paper to the underside of the rafters, thus forming air channels. The air is pulled directly to the heat pump using the roof space as a plenum. This increases the air temperature by between 3.5° F. and 6° F. — even in London which remains cloudy for most of the winter. The air is vented via the cowl or attic vent.

This system turns an ordinary slate or tile roof into a low temperature solar collector, and in most cases all it requires is paper, staples or tacks, and a cowl.

The packaged heat pump Keable uses is modified to a split, air-to-water unit. Refrigerant lines, made of copper, carry the superheated gas from the attic compressor to the basement where a water-to-water heat exchanger, supplied by Lennox, transfers the heat to a thermal store in the basement. Before heating the store, the gas is desuperheated to supply between 50 and 66 percent of

Better Homes and Gardens

Heat pump in the attic uses indirect passive solar gain plus heat that escapes through ceiling. Cooled air is blown out of house.

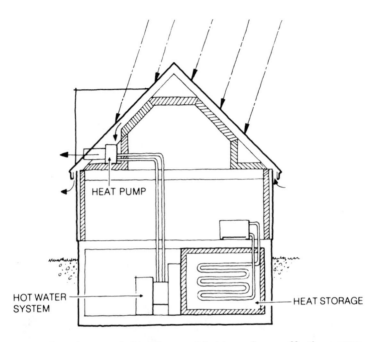

Heat pump in attic, storage in cellar combine to make an effective system. Note path of air across roof before it reaches the heat pump.

the domestic hot water needs of the house during the heating season. The desuperheater is connected with a preheating water cylinder.

The thermal store is housed in two separate PVC pool liner sacks sealed to form bags. The bags hold 2,600 gallons of water and are insulated with five inches of polyurethane sheet on top, four inches on the side, and one inch on the bottom. The store, which is two feet deep, covers a wide basement area.

Heat from the store is distributed to the house by hot water radiators.

Controls

The controls consist of a thermostat to switch off the system when the store reaches its set point of 120° F., an external thermostat set to switch off the heat pump at 30° F., and a time clock to switch the heat pump on in the morning and off in the evening. The heat pump does not sense the conditions in the house itself; it simply supplies the needs of the thermal store. The heat store supplies the house with background heat at 63° F. Additional heat is supplied as required in individual rooms by gas heaters.

The system was operational in January 1976. The cost of the heat pump, thermal store, heat exchangers, and other necessary equipment was $3,700. Add to this the cost of radiators and a platform over the thermal store in the basement to allow for storage space which cost $1,200. The internal insulation, double glazing, and draft-proof air locks cost $800, and the outside wall insulation cost $1,680.

Keable estimates the payback periods as three years for the insulation and seven to eight years for the heat pump and thermal store. These figures do not account for inflation or increased

Better Homes and Gardens

This conservatory, warm and bright with flowers during much of the year, provides extra heat to the house through direct passive solar gain.

energy costs. As energy prices have more than doubled since 1976, the payback period has been substantially reduced.

Additional energy features of the house are a conservatory outside the drawing room windows at the back of the house and a 200-watt Winco Wincharger aerogenerator mounted on the roof. The generator supplies twelve-volt lighting throughout the house using DC spotlights.

Parallel project

More recently, Keable and Dodson incorporated their ideas in a new owner-built house north of London. It is a frame house of lightweight construction, the kind that quickly loses heat. The walls are insulated with four inches of fiberglass and the windows are double glazed.

The heat pump draws air, preheated by solar insulation on the

Better Homes and Gardens

This home, conventional on the outside, has an attic heat pump, with an insulated plastic pool liner for storage of heat.

roof tiles, to the "outdoor" air coil in the attic. The heat in the superheated gas from the compressor is transferred to water in a 6,500-gallon water store beneath the house.

The store, made of a plastic pool liner, is insulated with two inches of expanded polystyrene at the sides and is topped with fiberglass insulation. There is still heat loss, but most of it tends to rise into the house. The water can store heat for ten days. The output of the heat pump is maintained at a steady 60° F. This is boosted by direct solar gain through the double-glazed windows, by recycling waste heat, and by incidental heat from lighting, appliances, and people. Initially the heat pump provided indoor heating at 70° F., but with all the incidental heat gains this was too hot, so the heat output was lowered to 60° F.

In summer, the house is kept cool by reversing the heat pump cycle to cool the heat store. The heat store is cooled during nighttime, when the outside air temperature is at its lowest and there is no solar gain on the roof. Since the heat store is in three separate "bags," one could keep two hot and one warm or vice versa to arrive at daytime cooling and nighttime heating.

Advantages

The innovative feature of this attic system is that it shows how simply slate, tile, tin, or any other heat-absorbing roof surface can be turned into a low temperature heat source at almost no cost. In addition, inevitable heat loss through the ceilings is not lost to the atmosphere but contributes to the heat pump.

By taking the outdoor coil from outside the house to the attic, one removes an eyesore and makes possible the enjoyment of a garden or patio without the background sounds of a noisy compressor and fan.

With a twelve-hour minimum thermal store, attic heat pumps are perfect for both heating and cooling. Where there is off-peak electricity available at night, summer cooling can be made even less expensive if the thermal store is charged at night for daytime use.

Finally, as an added bonus, attic installation removes the wind chill effect as the coil is protected from high winds and air turbulance. The "outside" coil is also protected from snow and freezing winter rain.

Disadvantage

The one disadvantage of this system is the potential for noise and vibration in the attic if the compressor is installed there. A properly designed, low-energy fan to blow the air over the attic coil will cause little or no irritating noise, but when a compressor starts it usually jumps and then vibrates as it runs. Compressor vibration is carried by the refrigerant lines. Furthermore, the compressor needs to be fixed to a solid mount such as timber beams, and this will carry the noise and vibration downstairs.

If the compressor is placed in the attic, the floor joists should be sound, the attic floor should be insulated, and the compressor should be mounted on rubber pads. Site the compressor above a bathroom or stairwell. The optimum location is downstairs next to the heat store where it is far less likely to cause sleepless nights. Vibration from a basement or ground floor will carry far less than from roof beams, and installation may be easier.

Both of these houses are undergoing extensive testing over a three-year period. It is hoped that the innovative work of Keable and Dodson will result in further development of a heat pump system specifically designed for attic installation and thermal storage.

SUMMARY

Location:	The Keable house, London
Date of completion:	1976
System:	Attic plenum. Air-to-water heat-only heat pump, wet radiator heating
Heat pump:	Lennox modified packaged system
Thermal store:	2,600 gallons water
House size:	2,700 sq. ft.
SPF:	2.8
Cost:	Heat pump, thermal store, and insulation — $3,700

Swimming Pool System: Air-to-Water

Some friends in Dublin, Ireland, were spending a small fortune heating their swimming pool with electricity until they installed the English-made heat pump Calorex, which has a COP of four at an air temperature of 55° F. and a COP of five at 75° F., when the pool is heated to 75° F. (a comfortable temperature for recreational swimmers). The heat pump plus a plastic air bubble pool cover for insulation combined to reduce heating costs by 90 percent.

During the season for swimming in outdoor pools, air temperatures are generally such that a COP of 4.5 can be expected. The high COP for summer heating is explained by the low temperature differences between the air and water; year-round indoor pools present more of a water-heating problem because of the low outdoor temperatures.

A heat pump provides inexpensive heat for this swimming pool.

The heat pump is fitted in a small shed, together with the pool filter and the old electric heater that was so exorbitantly expensive. The installation of the heat pump caused no problems. It is connected to the pool system after the water has passed through the filter, and the same pump that is used for filtration is also used to circulate the water through the heat pump. The evaporator for the heat pump faces north and is in a shaded spot, which is not the best for efficiency but does keep noise away from the swimming pool.

In more southern climates with a lot of sunshine, solar collectors for heating the swimming pool would be cheaper than a heat pump. With collectors, the initial cost is higher, but there are no further bills; with a heat pump there is always the operational cost of electricity.

Calorex also manufactures a large heat pump for an Olympic-

In a mild climate, this system works: solar collectors on roof heat the swimming pool. The pool then is the sole source of heat for the water-to-air heat pump system that provides the heat for his house.

sized swimming pool. This heat pump is directly driven by a gas-fired generator and has a COP reported to be ten. Both Westcorp and Phoenix have recently begun manufacturing heat pumps for swimming pool heating. Similar heat pumps can be used for industrial and commercial purposes; in laundries they would be particularly invaluable.

SUMMARY

Location:	Dublin, Ireland
Date of completion:	1978
System:	Air-to-water unit for swimming pool application
Heat pump:	Calorex (UK)
COP:	Between 4 and 5
Cost:	$2,500

Summer Ice, Winter Warmth: Air-to-Water

The following two systems are a contrast of complexity versus simplicity. The first system is the Knoxville House built by the International Energy Agency as a test installation to be monitored by Oak Ridge National Laboratory, Oak Ridge, Tennessee. The second is a house in Salford, England, built by the Salford Council.

Knoxville house

By the end of 1977, the cost of the Knoxville House was $373,000, which includes the cost of house construction and the data acquisition system. The project is being paid for by the Department of Energy (DOE).

The key to the system is 18,000 gallons of water stored in a tank. The theory of the Knoxville system is to store waste heat from the

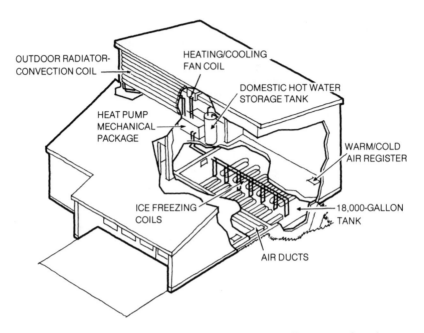

This house in Knoxville has an 18,000-gallon tank in cellar. Waste heat is stored there in summer, and heat pump extracts it in winter, converting the water to ice, which is then used for cooling in summer.

house in summer for winter warmth and to leave the water frozen at the end of the winter for summer cooling. It sounds like a tall story, that heat from air-conditioning could supply winter heating needs and ice produced in winter could supply summer cooling. However expensive, the system does work.

Ice-forming coils are horizontally and vertically spaced in the tank. An antifreeze solution passes through the coils to take heat from the tank to the heat pump exchanger. The heat pump package also contains a desuperheater, which takes the remaining heat out of the condenser and is, in effect, a mini-condenser which heats the domestic hot water. A radiant/convector coil is located on the south-facing vertical wall between roof sections to serve as a heat source or dump for balancing the system against year-to-year variations in climate.

During an unusually hot summer, the stored ice will be melted before the end of the cooling season, in which case the nighttime operation of the heat pump refreezes the water for the next day's cooling. This makes use of the cooler night air for disposing of waste heat and shifts the electric load from peak to off-peak hours.

Results

The house is unoccupied but contains a load imitation package that simulates the activities of a family of four. The tests show the following comparisons:

TYPE OF SYSTEM	ANNUAL SYSTEM SPF
Electric space and water heating, air conditioning	1.2
Heat pump, electric water heating	1.5
Annual Cycle Energy system	3.6

Advantage
- High SPF

Disadvantages
- Only DOE with its large budget could afford this system.
- Complexity
- 18,000 gallons of water weigh 150,000 pounds, which could be dangerous underneath a house in an earthquake zone.

Salford House

Annual energy cycles are beyond practical domestic applications; daily energy cycles are of more immediate use and are more economical as shown by the Salford House.

The Salford House is heated for less than 15 percent of what it would cost for a comparable house because of a combination of heat pump and tight insulation. The house, a small, two-story, two-bedroom unit, has insulation in the walls and attic as well as beneath the basement floor, double glazing, and a heat recovery unit on the ventilation duct.

Indoor temperature is maintained at 71° F. with the bedrooms at 69° F. Heat is provided by what must be the most inexpensive system in the world — plastic garden hose embedded in the concrete floors. Water flows through the hose at 87° F. There is no need for radiators, air ducts, or any external evidence of the heating system.

The house uses three small heat pumps made of components from domestic refrigerators; one is for domestic water heating and the other two are for space heating. They extract heat from a metal water storage tank and operate using off-peak nighttime electricity. By morning, water in the tanks has lost so much of its heat that it has frozen. Once the storage tank has discharged its heat, the house itself acts as a giant storage heater. Ventilation air — one complete air change every hour — is passed over the tanks of ice to melt them. Most of the heat normally lost to ventilation is recovered this way and by evening the tanks are melted for the heat pumps to switch on again. Additional heat to make up for natural losses can be gained from air or water sources.

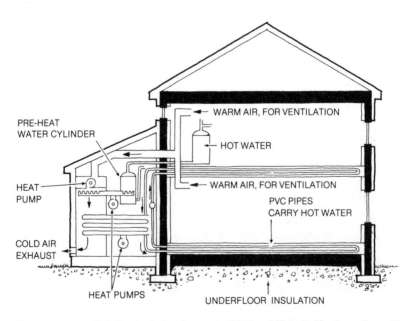

Heat pumps and tight insulation have cut heating bill in Salford to about 15 percent of what it would cost for a conventional system.

One pump alone fitted with a desuperheater could replace the three used in this installation and thereby reduce capital cost without any loss of efficiency. If this were done, the cost for a small two-ton (24,000 Btu) heat pump should not exceed $2,000 installed. The garden hose might cost $500 at most, and as floors are required anyway, they are not counted as part of the cost of the heating system. The water storage tank, being of metal, costs $1,200; if it were a plastic pool liner, the cost would be less than $500 with insulation. Ventilation heat recovery units are available on the market today and cost a few hundred dollars for small domestic units.

Comparison

The Knoxville House cost $373,000, operates on an annual cycle, has an 18,000-gallon thermal storage tank, and is unlikely to meet with broad consumer acceptance.

The Salford House has a 1,200-gallon thermal store, can use a standard heat pump (as can the Knoxville House), has what may well prove to be the most inexpensive and inconspicuous heating system ever used with a heat pump, and is clearly applicable to the broad range of new housing today.

Insulation, heat recovery, and the use of heat pumps have the combined potential for reducing home energy costs by as much as 80 to 90 percent. The Salford House shows one approach to achieving this goal inexpensively.

Heat Pump Water Heaters

Heat pumps made solely to heat domestic water are fast gaining popularity. In 1981 only 50,000 heat pump water heaters had been installed in American homes, but the thirteen manufacturers — nine more than in 1980 — predict that annual sales will reach three million units by 1990.

These heat pumps were first offered for sale in the 1950s, but never caught on, primarily because energy was cheap.

Now, with the public acutely conscious of fuel costs, homeowners are reading with hope the tests by both government and industry that show savings of 35 to 60 percent in homes that have used electric or oil-fired water heaters. The units cost from $800 to $1,200 installed.

Robert Hill, president of E-Tech of Atlanta, was quoted by the *Wall Street Journal* as saying that a family of four paying six cents a kilowatt-hour for electricity "will reduce the cost of heating water by 50 percent, saving about $200 a year."

These tiny heat-only air-to-water heat pumps absorb heat from *indoor* air and transfer it to the domestic water tank. Taking the heat from indoors is fine in summer if the air temperature is too high, but costly in winter unless waste heat is used.

Because heat moves from hot to cold areas, there is always waste heat in basements and crawl spaces. It may be waste heat from a basement furnace or laundry room; it may be heat usually lost to the ground below or lost through the cellar walls; it may be heat

E-Tech's heat pump for domestic hot water heating.

passing down through the basement ceiling. It is this heat that is ideal for the heat pump water heater.

These units take air at 45° F. or higher and supply hot water at 110° to 140° F. Under ideal conditions, 68° F. air temperature and 70° F. inlet water temperature, these units operate with a COP of three.

It is pointless to install one of these units if:

- You use a heat pump or air conditioner and have not installed a desuperheater.
- The source of heat for the heat pump is electricity or oil.
- You have not fully insulated your house against heat loss.

Desuperheaters may be a better choice. They cost as little as $100 in components for the handyman, or as much as $600 installed. On an air conditioner they reduce annual water heating costs by 50 percent; on a heat pump, the figure is more like 75 percent. In addition, desuperheaters improve the efficiency of air conditioners and heat pumps by between 8 and 16 percent.

Heat pump water heaters are ideal for hotels, restaurant kitchens, and laundromats. These places have an excess of hot air, and a need for cooling and hot water. Instead of dumping excess heat into the atmosphere, the heat is contained within the building and usefully employed in providing hot water.

Heat pump water heaters are manufactured by:

E-Tech Inc.
3570 American Drive
Atlanta, GA 30341

Energy Utilization Systems Inc.
365 Plum Industrial Court
Pittsburgh, PA 15239

Water as a Heat Source

To heat one's home, office, or factory with what feels to be cold water is far from fantasy. In fact it is eminently practical where water temperatures range from 40° F. upward. Below 40° F. the outlet water temperature gets close to freezing and may clog the return pipeline with ice.

Nationwide sales of water-source heat pumps reached 70,000 units in 1979, up from 40,000 units in 1977. This equals 14 percent of the total heat pump market at present. Market research carried out by *Air-Conditioning, Heating and Refrigeration News* points to a strong continued growth of 30 percent to 35 percent annually. In Florida alone, where well water heat pump systems are nothing new, 40,000 units provide economical heating and cooling. These are almost all water-to-air systems.

There are three types of water suitable for heat pump use:

- Well water
- Surface water
- Sea water

Well water

The average annual temperature of well water is 50° F. Seasonal variations rarely go beyond plus or minus ten degrees because below the frost line temperatures remain quite constant. Well water systems are potentially available to as many as 85 percent of all American homes, according to the Water Well Association. For this reason and because of their high working temperature, water well heat pump manufacturers are energetically promoting their product.

Well water systems are discussed in greater detail in the next chapter.

Surface water

Surface water (lakes, streams, and ponds) is more readily available in some parts of America than well water, particularly so where deep or difficult well drilling is more expensive than pumping water from a lake or river to the point of use. The disadvantage with surface water is that it is more susceptible to fluctuations in air temperature than ground water. However, water under pressure, i.e., at the bottom of a lake, always remains above 32° F. unless the lake freezes.

Surface water is suitable with a heat pump where a water turbine is used. Once the water passes through the turbine it can give some

The water wheel in the millhouse on the left drives a heat pump that heats this large manor house and part of mill, in a water-to-water system.

of its heat to a heat pump and thereby increase the effective output of heat by a factor of three. Such an installation is described in the chapter entitled "The Watermill."

Sea water

Except in the Arctic and shallow inlets, the sea does not freeze. It forms a great thermal reservoir, heating up a little in summer and cooling down in winter, but remaining at an average of 50° F. All along our extensive coastline there are restaurants, hotels, and houses, many of which could be using the sea as a source of heat in winter and as a heat sink in summer.

Later we will look at how one homeowner tapped this source.

Operating temperatures

Water-source heat pumps for industrial waste heat utilization can be designed to work with various water-source temperatures, ranging from just above freezing to 100° F., but the heat pump should have a design temperature suitable for the water source available. In addition, the design on the indoor coil must also be suited to the home or industrial heating needs.

A selection of four manufacturers shows the temperature ranges of current models in the heating and cooling modes for water-source heat pumps.

MANUFACTURER	TEMPERATURE RANGE FOR WATER-SOURCE HEAT PUMPS AT EVAPORATOR COIL		
Command-Aire:	heat/cool:	60° F. —	100° F.
Florida Heat Pumps:	heating:	58° F. —	85° F.
	cooling:	45° F. —	95° F.
Friedrich			
Standard models:	heat/cool:	60° F. —	75° F.
Low temperature			
models:	heat/cool:	45° F. —	65° F.
Vanguard Energy Systems:	heating:	40° F. —	80° F.
	cooling:	45° F. —	85° F.

This is a Phoenix water-to-air heat pump. The water coil and compressor are in the lower section. The air coil and blower are in top section.

If the available water supply temperature is 50° F., buy a heat pump which will operate below 50° F. because there are times when severe weather will reduce the water temperature by a few degrees or more. A heat pump not adequately designed to work with temperature fluctuations in the water supply will overuse expensive resistance heating; correct design on the heating and cooling side is also important.

The efficiency of water-source heat pumps is generally far higher than air-source heat pumps. The average SPF of air units tends to be less than two, whereas the COP for water-source heat pumps is between 2.5 and 4, averaging between 3 and 3.5 if the water source is 60° F. and the heat pump output is 110° F. or lower.

The higher the water-source temperature and the lower the end use, the better the COP. If water is available, it will probably be more efficient to use than air. In addition, the even temperature of water contributes to the more stable operation of the heat pump, resulting in a more reliable, longer-lasting heat pump.

Air-source heat pump COP's include energy lost to operating fans at the evaporator and condenser. Water-source heat pump COP's do not include water pump losses on supply and output sides. If a fair comparison is to be made, these losses, which may reduce water-source COP's by 5 to 25 percent, should be considered.

Scaling and incrustation

Water usually contains micro-organisms, minerals, and, sometimes, sediment which clog the water-to-refrigerant heat exchanger line from the water source to the heat pump. Where a high mineral content is indicated, the heat pump buyer must insure that there are provisions for cleaning away incrustation because it will reduce heat transfer. Mineral content in some water can form a thick, hard crust which may need to be chiseled away.

With sea and surface water, there must be a provision for clearing out any growth or sediment in both the coil and water lines. The water line also should have a filter at the water entry point.

Water piping

Pipes from the water source to the heat pump should be buried below the frost line to prevent freezing.

When designing the piping system, a simple rule of thumb is a flow of three gpm (gallons per minute) per 12,000 Btu (or ton of refrigeration). The following are some typical pipe sizes for various flow rates:

FLOW	PIPE SIZE
2–5 gpm	½ inch
5–10 "	¾ "
10–20 "	1 "
20–30 "	1¼ "
30–40 "	1½ "

Advantages of water-source heat pumps
- Higher COP than air-source
- Lower running costs
- Increased life cycle
- Reduced maintenance

Disadvantages
- Scaling and incrustation in coil and pipelines
- Energy expended pumping water from the source to the heat pump. The amount of energy used this way depends upon how water must be pumped and how far.
- Pipelines laid to and from the heat pump
- In the case of well water, a well must be drilled.

A list of manufacturers of water-source heat pumps is included in the Manufacturers' Index.

Water Well Heat

Dr. Carl E. Neilson, professor of physics at Ohio State University, may have been the first person to design, build, and install a ground water heat pump system. Using standard refrigeration components, Neilson put together a small unit in 1948 to heat his cottage. The first unit did so well that he installed a second ground water heat pump when he built a new 1,000-square-foot home in 1955.

He drilled an eighty-foot well and found the 44° F. water had enough flow and heat to supply his needs. He ran the outflow to a small pond beside the house.

The heat pump supplies 12,000 Btu for the one kilowatt of electricity with a COP of 3.5, which is excellent for 44° F. water.

Water in Neilson's area is generally of poor quality, consequently corrosion and the growth of iron bacteria were expected. To cope with this, Neilson made his own water-to-refrigerant heat exchanger. He used a section of 1-inch diameter pipe and a 5/8-inch pipe, put one inside the other, and twisted them into a coil. It has worked fine. Neilson cleans out the heat coil every two or three years, but the expected corrosion and growth of iron bacteria never occurred, probably because of his choice of materials for the heat exchanger. Cupronickel was used for both pipes in the exchanger. Cupronickel has the ability to expand and contract sufficiently during heat cycles to inhibit the growth of scaling. At least three heat pump makers use cupronickel for their heat exchangers.

Battelle Institute

One of the first large-scale users of well water as a heat source was the Battelle Institute of Columbus, Ohio. In 1958, this scientific research institute was equipped with two heat pumps feeding off five sixteen-inch wells drilled to fifty feet. The water source is 54° F. and the average COP is 4.4, with one of the buildings having a COP of 5.4.

In a report, consulting engineers recommended that Battelle install the water well heat pump. They also recommended that fresh air be preheated in winter with the well water before the water is used by the heat pump. Individually controlled water coil heating and cooling units are located in each office. A small heat distribution fan is behind each coil which in turn is set at 95° F. for heating, and 45° F. for cooling. The fresh air supply is separate from this system.

Two 6-million Btu (500 ton) heat pumps are used to service a 317,000-square-foot section of the building. The heat pumps are modified centrifugal water chillers made by the Trane Company.

Heating and cooling cycles are determined by water flow valves which direct water from the closed loop serving the institute to the condenser for winter heating and to the evaporator for summer cooling. The water-to-water heat pump has a fixed refrigerant circuit.

The institute is equipped with oil-, gas-, and coal-fired steam plants, which have provided heating during the twenty-two years the heat pumps have been at Battelle. The cost figures for heating with these various fuels are shown on the following page.

FUEL	1977	1978
Heat pump	$2.27*	$2.70
Natural gas	2.54	3.00
Fuel oil	3.64	3.85
Coal	2.26	2.88
Electric resistance	7.35	8.76

*Per 10^6 Btu

The heat pumps were used continuously from installation to the winter of 1964–1965. In 1964, a new oil- and gas-fired steam plant was brought on line, and this proved a cheaper means of heating than the heat pump.

With the oil embargo in 1973 Battelle was placed on an allocation of natural gas, so by the winter of 1974–75 it became desirable once again to use the heat pump. In 1976, to further diversify its energy sources, Battelle converted one of its boilers in the steam plant to burn coal.

Some problems were encountered with the water well heat pumps. The piping was fouled with bacteria from the well water. The pipeline was cleaned and a chlorinator was installed. Corrosion and scaling of the heat exchanger tubing was another problem. This was solved by installing magnesium sacrificial anodes at the heat exchanges and adding 1.0 PPM of polyphosphonate to the water.

National Water Well Association

The true champion of water well heat pumps is the National Water Well Association (NWWA). The organization claims that 85 percent of the nation's homes could get all their heating and cooling from ground water heat pumps.

This non-profit research institute has an eighty-member staff, and publishes a broad array of material on this type heat pump, ranging from a sixteen-page brochure (25¢) and a free catalog to a detailed and personal analysis of an individual's home requirements. Other offerings include films, slides, and tapes. It sponsors many conferences throughout the United States.

The address is:

National Water Well Association
500 West Wilson Bridge Road
Worthington, OH 43085

The association estimates that the cost of drilling a well adds an average of $2,400 to the cost of the total heat pump system. On top of this is the cost of the heat pump, $2,000 to $4,000, depending on size. The resulting investment of $4,400 to $6,400 is sizable but the NWWA claims that the payback period is from four to eight years.

A water well heat pump needs a minimum flow of about 2.5 to 3 gallons per minute per 12,000 Btuh needed. The NWWA says that if one randomly drills to 200 feet anywhere in the United States, a flow of 3 gallons per minute or more will be found 80 percent of the time.

Return of water

Ideally, the water should be returned to the water source via a second well. This can be a bored hole filled with gravel to keep the sides from falling in. The hole should be two or three feet in diameter and drilled to about twenty or thirty feet. This type of discharge pit is inexpensive.

Other ways to return the water are to use a pond or lake, or to irrigate with it. If you irrigate, a free flow system is best as there is little pressure in heat pump water pumps. Florida may pass a law so that water well pump users will have to drill a second discharge well. Other states have environmental restrictions about returning heated water into discharge wells or bodies of water. Check your state laws before installing a water well heat pump.

With an average COP of 3.5 (compared to 2 for air heat pumps), water well heat pumps are certainly worth investigating. The key to the cost is how far one has to drill for a well; the less you have to drill, the better. One way to find out is to ask a local driller for an estimate, usually a free service.

Sea-Source System

In San Diego, California, I visited a house heated and cooled by the sea. The house is built on the edge of a harbor protected from the Pacific surf by a sand barrier. The biggest waves that the pipeline carrying the sea water to the heat pump had to contend with were those created by small pleasure boats.

A mixture of solar collectors, two 1,500-gallon storage tanks, a heat pump, four water pumps, and a makeshift control panel, combined to create a complex, unnecessarily expensive system.

Home locations near the sea provide owners with access to an inexpensive and constant source of heat for operating heat pumps.

Heat is transferred to and from the sea by means of a closed loop in which water is circulated by a centrifugal pump. The first piping installed was copper with brass connections. The brass fittings disintegrated in the sea water; the pipe was replaced with plastic but plastic is a poor conductor of heat. The plastic pipeline stretches alongside a small jetty. The end of the pipe is weighted down with concrete.

Heat transfer from the closed loop can go directly to the heat pump or to the water storage tanks which are placed underground beside the house. One tank is high temperature water for heating; the other is low temperature for cooling. Both tanks are uninsulated.

Solar collectors located on the flat roof feed the high temperature storage tank. When the storage tank is above 100° F., it supplies heat directly to the fan coil. When the temperature goes

The two plastic pipes next to the pier carry water from the sea to the house, where a heat pump extracts the warmth for heating the house.

These solar collectors provide backup heat for the house, but, considering their initial cost, are of dubious value.

below 100° F., the water is taken to the heat pump to be upgraded in temperature. The fan coil takes the heat out of a second water loop on the condenser side of the heat pump and distributes it to the air duct system in the house.

The second water storage tank, the low temperature one, is not used. The heat pump uses sea water directly for cooling.

The control system is spread over three electrical panels.

The total system, which cost $17,750, supplies heating and cooling to the house, and hot water for the Jacuzzi and domestic needs. State and federal tax credits reduced the cost to $11,000.

A far simpler and more direct system would be to use a water-to-air heat pump, transferring heat to and from the sea by a more efficient closed loop than plastic (possibly cupronickel). With sea water temperatures ranging from a low of 50° F. to a high of 70° F., COP's in the order of three and four can be obtained. A desuperheater could take all the waste heat during the heating and cooling cycles to heat at least 50 to 60 percent of the hot water and the Jacuzzi.

This system would eliminate the need for storage tanks, solar collectors, and three of the pumps, and would vastly simplify the control system. The sea is an excellent "storage tank," especially in southern California. The uninsulated storage tanks must lose heat, making them only marginally better than the sea. Once the heat pump is installed for a year or so and fuel costs are known, then a clearer decision can be made on the need for solar panels.

The cost for the simpler system would be $4,000 installed, using a Vanguard four-ton heat pump with the ability to operate down to 40° F. and a desuperheater. The cost includes the closed loop to the sea and a ½-hp submersible pump. The submersible pump would reduce noise. Retrofitting the house with air ducting would cost about $3,000.

Beyond improving the technology of the heat pump itself, nothing can improve the cooling cycle on the suggested simpler system. Heating domestic and Jacuzzi water with the heat thrown off by the condenser is perfect, and remaining heat can go to the sea. In the winter, the southern sea itself is an ample heat reservoir for home heating with a heat pump.

Advantages
The designers claim a COP of six to eight for the total system. However, there are no test figures to back this up.

Disadvantages
- High initial cost
- High maintenance costs due to complexity

Finally, the use of the sea as a heat source and sink requires a reasonably calm water inlet, lagoon, or harbor where buildings are close to the water's edge. A heat exchange coil on an exposed ocean front would be stressed too much, not just from high waves, but also from flotsam and jetsam.

SUMMARY

Location:	San Diego, California
Date of completion:	1977
System:	Sea-source water-to-water heat pump with air coil for duct heating
Heat pump:	Advanced Design Associates, Florida
Thermal Store:	Two 1,500-gallon water tanks
SPF:	Unknown, design COP of 6–8, unverified, cooling EER 12–15

Cost:		
	Heat pump	$ 3,000
	Storage tanks	2,000
	Solar collectors	2,750
	Air ducts	3,000
	Control system	1,600
	Pumps and valves	1,000
	Piping	400
	Labor	4,000
	Total	$17,750

The Watermill

I was driving with a heat pump manufacturer in England when we saw an old millhouse. We went inside, met the owner, and found that his house is completely heated by a water-driven heat pump. An old twenty-foot-diameter, breastshot water wheel operating on a ten-foot head and turning out about five or six horsepower (3.75–4.5 kilowatts), directly drives the compressor of a heat pump which takes the heat out of the river and transfers it to a wet radiator heating system.

The owner of the house took justified pride in showing me the installation, which was built by a friend of his, Ellis Robinson, an avid heat pump enthusiast. The system could almost be called stark in its simplicity. The V-belt-driven compressor, refrigerant lines, and associated equipment are the work of a fine craftsman.

A water-to-refrigerant coil in the tailrace just by the water wheel absorbs heat from the river, which varies in temperature during the winter heating season from the low 40's to 50° F. This is a heat-only system. Cooling is unnecessary, so the river coil is a fixed evaporator. As the river flow is approximately 700 cubic feet a minute, the temperature of the water crossing the evaporator is reduced by a fraction of a degree. Because the temperature loss across the coil is so low, evaporation is more efficient and, of course, there is no risk of ice forming on the coil.

The compressor is mechanically powered by the water wheel which means the need for a generator on the water wheel and a motor on the compressor is eliminated. Water turbine generators and small electrical motors have efficiencies of about 80 percent. By not using electricity, this system is 36 percent more efficient and less costly to install than motors and generators. The speed of the compressor can be adjusted mechanically. The output of heat in

At lower right is the compressor. It is driven by mechanical power directly from the water wheel.

At lower left is the drive from the water wheel. Compressor is at left in rear. Both power and heat for system come from the river.

the home can be adjusted by cock valves on each of the water-filled radiator panels.

The fixed condenser is a refrigerant-to-water coil which supplies domestic hot water and heats the closed loop for the radiators. The most obvious indication of the success of this installation is to feel how cold the water in the stream is and then to feel the temperature of the condenser water at 110° F. to 120° F. The gas in the refrigerant line from the compressor to the condenser is closer to 200° F.

The U.S. Army Corps of Engineers has estimated that if all the existing dams in the United States plus a few ideal hydro locations were used to generate electricity, the resulting output would equal 25 percent of the total electricity generated in this country. The use of water-powered heat pumps can effectively triple and even quadruple the energy gain in heat for large or small factories, offices, and homes that use water turbines.

If the house in this installation were to depend for heating on the

This river turns the water wheel in mill at left, providing heat for house.

output of the water wheel alone, it would be a cold house. The heat pump transforms this, multiplies it by a factor of 3.5 or 4 to provide a warm and comfortable home.

Advantages
- No heating costs
- Easy maintenance due to simplicity
- Long life

Disadvantages
- There are none

SUMMARY

Location:	England
System:	Water wheel powered, water-to-water, heat-only heat pump
Heat pump:	Homemade, mechanically driven
COP:	Estimated at in excess of 3.5
Cost:	To replace the heat pump with a motor-driven, factory-made unit would cost more than $4,000.

Solar
Heat Pump Systems

A heat pump needs energy, usually between a quarter and a half of its output, for its operation. Solar energy requires considerably less. In fact, a totally *passive* solar house requires no purchased energy; even active systems such as an *active* solar water heater require only sufficient energy to operate the pumps or fans, plus a minute amount of energy to operate the temperature differential control. Passive solar buildings, if properly designed, cost little in addition to what one would normally pay for a house, and can supply anything from 10 to 90 percent of the domestic thermal energy requirement. If you are building and have a suitable site, choose a passive house design. If you are retrofitting an existing building, think about an active solar system. Solar collectors will supply energy in amounts dependent upon the size and number of collector panels and the storage that one installs. Open the house up as much as possible toward the sun and then make up the balance of heating and cooling requirements with a heat pump or possibly with a wood or coal stove.

Passive solar energy will cost between 2 and 5 percent of the construction cost of the building. (These percentage figures only apply to new buildings and include thermal storage costs.) For a home with a construction cost of $50,000, 2 percent is $1,000 and 5 percent is $2,500. With a heat pump costing an average of $2,500 to $3,000, the total figure ranges from $3,500 to $5,500. The cost of retrofitting a house for passive solar gain depends to a great extent on the structure and orientation of the house.

An active solar system plus heat pump is expensive. The solar panels, thermal storage, and heat pump cost between $7,000 and $15,000 — $15,000 is more the average. A Thomason trickle type collector is the exception: it costs much the same as the roof it replaces.

Systems

There are four main solar heat pump systems:

- *Passive solar heat pump* is a passive solar house with a separate heat pump for auxiliary temperature control. Or it is simply the passive principle applied to the heat pump itself.
- *Solar evaporator coil* transforms the outdoor evaporator coil into a glazed or unglazed solar collector.
- *Solar-assisted heat pump* uses solar collectors and heat pumps in either series or parallel.
- *Solar-powered heat pump* is driven directly or indirectly by solar heat.

The marriage of heat pumps and solar collectors is thought by some to have been ordained in heaven. The biggest problem with air-source heat pumps is low performance in freezing weather. Solar preheats the air or water source for the evaporator, raising the COP from 1.5 to as high as 4 and over. This is the lure of solar heat pumps — the promise of a higher COP while still using standard heat pumps.

Passive solar heat pump

Passive solar designs work with the arc of the sun which is low in winter and high in summer. In winter the low sun shines on the south side of a building and can be collected to heat the air within. This winter solar energy passes through glass which, because of the *greenhouse effect*, traps the solar radiation.

In summer the maximum impact of the sun is on the roof, and the east and west sides of the building. Installing exterior overhangs to shade glazed areas can reduce summer cooling by 50 to 90 percent; cross-ventilation and shady plants also will provide additional cooling.

These same principles can be applied to the outside heat exchange coil of heat pumps. The coil should be open to the warming and defrosting effect of the low winter sun and completely protected from unwanted summer solar gain. Glazing on the south side of the coil will increase the useful solar gain in winter, just as shade will decrease the useless summer gain.

There are few locations where the application of at least a partially passive approach is not possible for heat pumps. Between 70 and 90 percent of all new construction and existing homes with heat pumps could use a passive solar shell on the outside coil. The coil can be located on the ground a garage roof, or on the main roof of the building, especially if the roof is flat.

Advantages of passive solar shells
- No moving parts
- Low cost relative to heat gain
- Increased heat pump COP
- No wind chill
- Fewer defrost cycles
- Less auxiliary heat required
- Easy to install
- Can be homemade
- Excellent application of passive solar principles

Disadvantages
- None other than that it takes up a little room

Solar evaporator coil

A solar panel is a layer or two of metal with pipes or channels through which the fluid flows to collect heat. The panel is usually glazed, boxed, and insulated. Panels for low temperature use are sometimes made of plastic and unglazed, especially those for swimming pool heating.

A heat pump is an indirect solar collector in that its heat source is warmed by the sun. Put the outdoor evaporator coil of an air-source heat pump in an insulated box, glaze it, and leave it open to the sun, and you have a solar collector. Add a compressor (it can be a tiny one), a condenser (just like the heat exchange coil on a solar system), and an expansion valve, and that completes the type of solar heat pump that may revolutionize the heat pump industry; it is also the type of heat pump being worked on by Solartherm Manufacturing Corp. in Palm Springs, California.

"The Energy Machine" made by Solartherm is a refrigerant-charged solar collector, feeding a refrigerant-charged valance heating and cooling system. The collectors and valance are part of a closed refrigerant loop which includes a compressor and an expansion valve.

Solartherm claims that its "energy saving package" is so quiet and compact that it can easily fit in one's bedroom and yet not disturb sleep. The system is reversible; through the collector it can extract heat from the sun or outdoor air, and in the cooling mode it reverses the cycle and pumps heat out.

Most heat pumps will not perform when the air temperature goes below 10° F. to 20° F. except by means of auxiliary resistance heat. Solartherm claims its unit will heat down to 0° F. When air temperatures in summer soar over 100° F., the EER of heat pumps in the cooling mode falls dramatically. Some fail to operate at 110° F. or more outdoors, which is the time when cooling is most needed. The Solartherm unit cools at temperatures well over 100° F. The manufacturers claim a COP of 4.8, which is excellent, and an EER of 16.1, which is above average.

The valance, a finned coil, is the condenser in the heating mode and evaporator in the cooling mode. Room air is free to flow or cycle over the valance which is located near the ceiling of the rooms to be served. Heating and cooling are achieved by a mixture of radiance and convection. Condensation on the fins is used to effect air filtration; a condensate drain is part of the system. Because there are no fans or blowers, the system is noiseless, similar to a hydronic radiator.

All that happens in the valance is the evaporation and condensation of refrigerant. An average of sixteen square feet of valance coil area per 12,000 Btu (or ton of refrigerant effect) enables a much better heat transfer capability than the one square foot per ton on standard heat pumps. In addition a desuperheating kit is a standard item with this system.

Advantages
The advantage of using a solar evaporator coil is that it offers a high source of evaporator heat. This leads to exceptionally high COP's, far higher than anything else, such as water, earth, or air.

Solar evaporator coils qualify for federal solar tax credit. This tax credit alone, combined with local state credits, could be the biggest single factor to spur the development and marketing of solar heat pumps.

Disadvantages
The sun only shines during the day. Heat absorption can continue in the night, but not very effectively and certainly not if the solar coil is glazed and insulated. Therefore some type of heat storage may be needed for overnight heating.

The sun does not shine all winter. However solar collectors continue to be effective even in diffuse winter sunlight.

Cooling EER's will be low if solar panels are used for heat dissipation during the day.

Manufacturer
Solartherm Manufacturing Corp.
768 Vella Road
Palm Springs, CA 92262

Solar-assisted heat pump

The solar-assisted heat pump (SAHP) package consists of a solar panel with storage and a heat pump. The solar panel system is designed to heat the home; it is far bigger than solar panels used to supply domestic hot water. The heat pump is designed to heat

The solar collectors feed heat to heat pump, while greenhouse provides direct heat to this house. Tight insulation completes this energy package.

the whole house, using air or water as the heat source when the solar heat falls below a working temperature.

Because solar space heating and heat pumps are the two most expensive heating systems available today, the combination of the two can cost as much as $15,000 for a house of 2,000 square feet. Even with solar tax credits the payback period is long. A complex system requires careful planning to ensure that the component parts work in harmony; if not, the repair and servicing bills may be high.

This system uses either air- or water-source heat pumps. If the solar collectors and heat storage use air, the heat pump will be an air-source model. If the solar and storage use water, a water-to-air heat exchanger can feed an air-source unit, or even simpler, a water-source heat pump can work directly with the solar heat store.

Here is how a typical SAHP works:

- When the solar storage temperature is 100° F. or over, heat from the storage goes directly to the house heating system.
- If the solar storage temperature is between 50° F. and 100° F., and greater than any secondary heat source such as outside air or water, the storage becomes the heat source for the heat pump.
- If the secondary heat source temperature is above that in the solar storage and still within the operating temperature range of the heat pump, the secondary heat source is used.
- If neither the secondary nor the solar storage temperature is high enough, auxiliary electric heating is switched on automatically.

The system outlined is a series SAHP because the heat pump works with lower temperature solar heat. A different method is called a *parallel* system, with the heat pump working as an add-on

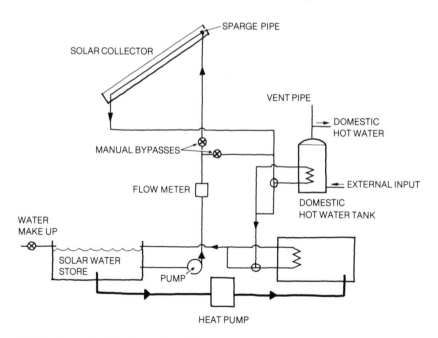

Design for a simple solar-assisted heat pump system.

to the solar system. The heat pump does not work with solar preheated air or water but is simply a conventional system. The July 1978 issue of *Solar Age* magazine argues that there is little difference in efficiency between the two systems.

For summer cooling the solar panels are of little use, and so cooling is achieved using air or water as the heat sink. In some systems the solar storage is chilled at night using lower outdoor temperatures and off-peak electricity. The disadvantage of this is that if the solar storage is used for summer cooling, it cannot provide domestic hot water without the addition of a separate, smaller tank.

The advantage of SAHP systems may be illustrated as follows: Outdoor air temperature is 15° F., the point at which most air-source heat pumps are clogging up with ice. Defrosting is then required, and the auxiliary heating is automatically turned on. However, the winter sun shining on the solar collectors heats the air to 55° F., still far below the 100° F. required for solar space heating, but high enough for an air-source heat pump to get a COP of three. As a result the heat pump can increase its output threefold, requires no defrosting, and is spared the extra wear and tear of having to operate at low ambient air temperatures.

There are advantages on the solar side as well. When the temperature difference between outside air and solar-collected heat is low, the efficiency is high, but when the temperature difference is high, the efficiency plummets.

If the solar panels were to provide heated air at 100° F. with an outside temperature of 15° F., the efficiency of the panels would be 40 percent, because of the 85° F. temperature difference. By using a heat pump with 55° F. air, the efficiency of the collectors increases to 65 percent.

Roots

The first commercial applications of solar heat pumps were made in 1954 and 1956 by the firm of Bridges and Paxton. The firm was awarded a contract to engineer the heating and air conditioning for a building in Albuquerque, New Mexico, that had glass walls on both the north and south sides. Solar gain on the south side resulted in overheating even on freezing winter days. Instead of buying gas to heat the north side and air conditioning to cool the south, the firm used the excess heat from the south to heat the north, thereby effecting a balanced temperature throughout. This is basically a passive solar heat system. In addition, a well was drilled as an auxiliary heat source/sink for the heat pump.

Advantages
- One test by Dr. E. Kash at Brookhaven National Laboratory showed that the solar temperatures when supplied to the evaporator of a heat pump (shown on the left column) provided a condenser temperature of 120° F. with COP's as shown on the right column.

SOLAR TEMPERATURE	COP
60° F.	3.8
70° F.	4.4
80° F.	5.0
90° F.	5.8
100° F.	6.5
110° F.	7.4

- Lowered annual energy costs
- With storage, can operate using off-peak electricity

Disadvantages
- In many cases the high cost of a SAHP results in an unacceptably long payback period
- Complexity, possibly leading to high maintenance costs
- Solar collectors are least efficient in cloudy weather; heat pumps are least efficient in extreme cold

The future
The future of SAHP may not be with high-performance, high-cost solar collectors designed primarily for supplying domestic space heating and hot water at temperatures of at least 100° F. Rather, their potential is with low-cost, low-performance solar preheating coils, such as plastic pipe or passive solar shells, where the capital cost is reduced considerably.

NASA engineers at the Langley Research Center in Virginia

built a house with a SAHP, 384 square feet of collector, a 1,900-gallon storage tank, and a water-source heat pump with well water back-up. They concluded that it is far cheaper to run the heat pump on well water alone, and that if they were to repeat the installation they would dispense with the solar panels, storage, and all the controls that go with them. Instead of the SAHP, they would use the simpler well water system for heating and cooling.

Solar heat pumps require more research and development than just putting together conventional solar panels and heat pumps. The answer, I believe, lies in passive solar preheating systems and in refrigerant-charged solar panels as part of a heat pump cycle.

Manufacturers

The following companies manufacture or design SAHP systems:

Airtemp Corporation

Carrier Air Conditioning

Fedders

General Electric

Phoenix

Vanguard

Westinghouse Electric Corp.

Solar-powered heat pump

Indirect solar power such as wind and hydro power can be used to drive heat pumps; even a heat pump driven solely by sun power is possible.

High temperature solar collectors can be used to evaporate refrigerant, and the high pressure vapor can power a turbine to drive the compressor of a heat pump. It is possible, but costly. AiResearch Co., 2525 W. 190th St., Torrance, CA 90504, has built prototypes on this concept under a government contract.

Rox International, 2640 Hideen Lake Drive, Sarasota, FL 33577, manufactures the Minto Freon engine and generator. The power generating units use solar, geothermal, or industrial heat over 150° F. to evaporate the Freon refrigerant to power the turbine generator. The mechanical power can be used to drive heat pumps and electrical generators.

A solar-assisted heat pump system in Lake Montclair, Virginia.

A similar system in Birmingham, Alabama, Both are by Westinghouse.

Photovoltaic cells, which generate electricity from the sun, may have a part to play with heat pumps, but not until the cost of the cells is greatly reduced. They could conceivably generate DC electricity which would drive a DC motor to operate the compressor.

The disadvantages of sun-powered heat pumps is that the technology is not yet developed and may prove expensive. In addition, they are dependent on the sun.

Solar-Assisted Heat Pump: Water-to-Water

Macclesfield, suburb of Manchester in central England, is the home of a unique experiment; an old, ramshackle coach house was transformed into a solar-heated home. The principal energy ingredients are insulation, solar roof, solar greenhouse, and a heat pump.

The structure was gutted and completely rebuilt. The floor was replaced with four inches of concrete over two inches of a load-bearing polystyrene insulation.

As the house had solid brick walls, insulation was added to the exterior of the house. This was done by running 2 × 2 timber battens horizontally and vertically, with insulation in between. The result is nearly four inches of insulation, covered with "breather paper" and then with weather-treated lumber for the outside skin.

The north slope of the roof is insulated with almost six inches of polystyrene; the south slope contains a Thomason trickle type solar collector. The collector is made of corrugated aluminum with the channels formed by the corrugations running down the slope of the roof. From a header pipe at the top, water trickles down the open channels and is collected at the bottom in a "solar gutter." A single sheet of glazing is supported by a glazing bar. The aluminum collector itself is supported by steel Z-purlins and is painted matte black. Thin sheets of asbestos were placed between the aluminum and steel to prevent corrosion.

The solar collector supplies both domestic hot water and space heating. Water is drawn from a 500-gallon solar water tank and pumped up to the trickle collector. From there the heat is transferred first to the domestic hot water tank, then to the space heating tank, and finally it returns to the 500-gallon tank.

The main space heating storage tank has a capacity of 800 gallons and is maintained at temperatures of between 105° F. and 115° F. The smaller 500-gallon solar water tank, which is used to feed the collector with water, is kept at between 40° F. and 80° F. If, after days of cloud cover during which the solar collector is not activated, the space heating tank falls below its design temperature of 105° F., then a two-ton, water-to-water heat pump is automatically activated to transfer heat from the solar tank to the space heating tank. That is the sole function of the heat pump.

When the 500 gallons of water (and anti-freeze) in the solar tank is reduced from 80° F. to 40° F., the heat pump will have removed 166,000 Btu, the equivalent of 48.6 kwh, which can then be used for home heating.

In the rare event of extreme prolonged bad weather, a 3.5-kw boiler automatically feeds a small high-temperature space heating tank.

The heat pump was designed and made on a one-only basis by Prestcold for $2,000. At the time, no manufacturer in England made a water-to-water heat pump.

There are today in America a number of water-to-water heat pumps suitable for this type of installation. What is required is a packaged unit which is installed by connecting two input/output water circuits from the solar water store and the space heating tank. The cost for a one- or two-ton heat pump for a similar installation — a well-insulated, solar-heated, 1,800-square-foot house — would range from $1,000 to $1,500.

An additional feature of this house is the conservatory or "lean-to solar greenhouse" built onto the southern front of the house. It features a rock pit for storing heat. During the day, warm air is drawn from the top of the greenhouse, down a duct, and over the stones. By night the stones have gained heat which can then be used to keep plants warm and prevent freezing. In addition, heat from the greenhouse can be drawn into the three upstairs bedrooms by means of vents at the top of the greenhouse and bottom of the bedroom windows. All that is required is a small fan.

Having spent more than a year living in the house, both Jeff and Lynn Grant are well pleased with both the performance and comfort of their reborn coach house.

SUMMARY

Location:	Macclesfield, England
Date of completion:	1979
System:	Packaged water-to-water using solar preheated water
Heat pump:	One-only design by Prestcold
SPF:	In the region of 3
Cost:	$2,000 in England. Standard packaged U.S. machines less expensive.

Pools and Pumps

It makes sense to use a heat pump for swimming pool heating — and even cooling — where there is insufficient space for solar panels or in a cloudy climate, but it is quite another thing to use a swimming pool as the heat source/sink for a heat pump system. The system design and heat load calculations have to be fine tuned to the pool size, climatic conditions, and house size. Changes in climate leave very little room for flexibility. Because swimming pools are usually small, the pool temperature falls quickly to freezing in a hard winter and can hit the nineties in summer. Such extreme temperature swings defeat the purpose of a pool for both heating and cooling.

Whether swimming pool heat pump systems are any better than high efficiency air-to-air units is a debatable point. One system that does work is a pool specifically built as an insulated energy store with a small heat pump used to serve the low energy requirements of an energy-efficient, preferably passive solar house. In such cases there is lots of leeway, as energy demand is well balanced with supply.

The supply of energy for a pool comes from two sources: direct sunlight and the earth. Direct sun shines on an insulating pool cover that acts as a crude solar collector — the most inexpensive form is a plastic air bubble. Should the pool fall below ground temperature in winter, there will be some small transference of heat — a similar cooling influence is gained in summer. Both of these heat gains require careful calculation if the system is to work.

The following two systems give an indication of the problems encountered by those brave souls who were first to dip into the swimming pool/heat pump systems.

Solar pool system

I found an example of solar energy used with a heat pump and a swimming pool at a single-story home in southern California. The house has an area of about 1,700 square feet. On top of the roof are 200 square feet of unglazed copper solar collectors. They directly heat the 15,000 gallons of swimming pool water. The old gas boiler previously used for heating the pool is still available if additional heat is required.

The system uses a water-to-air heat pump manufactured by Mammoth. In the heating mode heat is extracted from the swimming pool and passed directly over the evaporator of the heat pump, then the cooled water is returned to the pool where, on sunny days, it is reheated by the solar collectors. The heat pump has a COP of three when the pool water is 70° F. Should the temperature of the pool fall below 52° F., electrical resistance heating is automatically switched on. No figures were provided by the manufacturers for the COP at other pool temperatures.

The pump is situated in a narrow hallway adjacent to a bedroom and the main living room. It is noisy and, despite insulation, the owners find that they have to turn up the television or speak more loudly when the heat pump is in operation. The heat pump should have been designed so that it wouldn't make the noise it does, and it should have been located where it would not be troublesome.

Another error in design of this installation is that the only way the heat pump can operate is by drawing heat from the swimming pool, and the only way the swimming pool can be heated is either

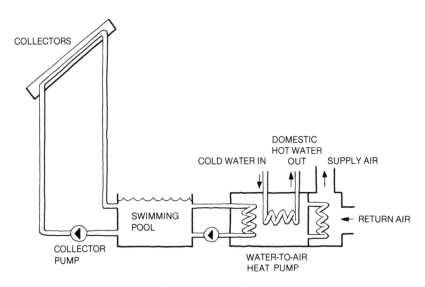

A Vanguard solar system with a water-to-air heat pump.

by solar collectors or directly from the gas boiler. In winter, to heat the swimming pool with the gas boiler and then take the heat out of the pool to the heat pumps is a roundabout way to approach home heating. It would be better to have the gas boiler heat go straight to the fan coil. If the pool were well insulated there would be less heat loss en route from the gas boiler through the pool to the heat pump.

A swimming pool that is not used in winter provides an excellent heat storage unit provided it is properly insulated with a pool cover. In most climates it is unnecessary to insulate the sides and bottom of the pool because the earth provides useful heat gain.

In the cooling mode the pump dumps heat into the swimming pool. Nobody has complained about the pool overheating in summer.

This type of installation in a retrofit situation is costly. Heat pumps are expensive, as are solar collectors, particularly if they are metal.

Needless to say, this installation is only of interest to somebody who already has a swimming pool or who is thinking of installing one. Choose the type of solar collector and heat pump carefully. If the manufacturer of the heat pump cannot give adequate or efficient performance figures, don't buy it.

If I were to install a similar system, first I would buy a unit that could work efficiently with water temperatures of 40° F. Second, the pool should be equipped with an insulating cover. Third, the auxiliary heat would be supplied directly to the air duct system and not to the heat pump via the pool.

The costs for a similar installation using a more appropriate heat pump are as follows:

Solar collectors (unglazed)	$2,500
Heat pump (made by Phoenix)	3,300
Pumps	400
Plumbing	600
Valves	800
Pool insulation	500
	$8,100

Installation costs for southern California are included in these figures; they assume that one has a swimming pool. Costs elsewhere may vary.

Advantages
- More use is made of the swimming pool
- Higher heat pump COP and EER

Disadvantages
- High cost
- Pool overheating in summer. This can be countered by using a cooling fountain.
- High maintenance costs

Alternative systems

Howard Barnhard was one of the first six people in the United States to build a house using a heat pump/swimming pool system.

His three-bedroom house in Arkansas, completed in 1979, is well insulated and uses wood as a back-up source of heat. The south-facing slope of the house is angled to take solar panels or photovoltaic cells at a later date.

The three-ton, water-to-air heat pump was designed and built by Advanced Design Associates who have since been taken over by Tempmaster International. The pool holds 28,000 gallons of water and is twenty feet by forty feet.

Summer cooling is achieved by dissipating heat through a small fountain into the pool. Winter heating is stored in and collected by the water in the pool, which is then pumped to the heat pump.

The system was designed so that additional resistance heating would not be required. However, since its installation the system has been under severe strain because of changing weather patterns which produced a harsh winter in 1979 and extreme drought conditions in 1980. Electric resistance strip heating had to be used in the winter, and the excessive demand for summer cooling increased the pool temperature to 90° F. and over.

During the winter of 1979, pool temperatures dropped to the low 40's early in the heating season. The heat pump continued to function at temperatures down to 35° F., albeit at a low COP, but just how low the manufacturer does not specify. Below 35° F. strip heaters automatically switch on.

As there are no solar panels involved in this system, the pool collects heat by direct solar gain through a plastic air bubble insulating cover. When temperatures drop below 55° F., some heat will be gained from the clay soil in which the pool is contained. Nighttime heat losses are considerable, due to the poor insulation. To improve the efficiency of this system, a heavier insulation material should be drawn over the pool at night.

In summer the cooling efficiency would be improved by passing the water from the heat pump over a small shaded rock water garden before it returns to the pool. Both summer cooling and winter heating could be improved by the use of buried heat pipes. This would conduct coolness to the rocks in summer and heat to the pool in winter. However, placement of heat pipe rods in the base of the pool could interfere with swimming. The cost of the heat pipes and installation could come close to that of a small well which would provide a better source of heating and cooling than the pool currently does under changing climatic conditions.

One advantage of this system, which costs in excess of $3,000, is that it has a desuperheater which provides most of the hot water requirements.

Asked if he would repeat the system, Barnhard is hesitant. He would reassess all heating options, would look closely at new air-to-air units, but more specifically, he would look at well water heat pump systems — and such a system could be tied in with the pool to give a far more effective heating and cooling system than he has now.

Earth Energy

A house heated by the earth must be one of the most satisfying heating systems; heat pumps make this feasible and practical. The depth of the frost line can vary from a few inches to four or five feet, but below this depth the earth maintains a fairly constant year-round temperature of between 45° F. and 75° F. True, there are seasonal variations, but rarely more than ten to twenty degrees Fahrenheit. Heat removed from the earth is replaced by natural conduction of heat in the earth. A heat pump can use the earth as both a heat source and a heat dump. As such, an earth-based heat pump system can function in the heating and cooling mode.

One large expense of an earth-source heat pump system is for placing the pipe underneath the soil. Copper or plastic can be used, and the heat exchange medium may be water or even refrigerant. Water with antifreeze is usually used and has to be placed in pipelines below the frost line. The pipes are placed in the bottom of channels which are then filled with either sand or loam. Plastic has much poorer heat conductivity than copper. As such, almost twice as much exterior surface of the plastic pipeline will be required.

The actual area of ground that needs to be used for pipeline equals approximately the square footage of the house and certainly no less than half. The length of piping required for a house of approximately 2,500 square feet would be up to 500 feet of copper tube and between 800 and 1,000 feet of plastic tube. These are approximate figures for temperate zones.

The surface contact is really what is important. Simply to put a water tank underground and expect it to absorb sufficient thermal gain to heat one's home is incorrect. The surface contact from the buried water tank is minute compared to the contact one would get with pipeline containing a similar quantity of water.

There are ways of increasing the heat gain from the soil: first by using solar energy, and second by using septic laterals. A septic lateral will carry the waste water into the soil. As there is heat contained in this waste, it will increase the overall efficiency of the heat pump system.

Both leach lines from a septic tank system and earth coil pipes were installed at the same time for new house, to cut excavation costs.

An ideal place to put the coils for an earth-based heat pump system is underneath a vegetable garden. In winter there is very little growth and so the dark surface of the soil is exposed to the sun, which aids its solar absorption. During the summer, the foliage of one's vegetables decreases the solar gain to the soil, thus increasing the ability of the earth to absorb heat from the house.

Another possibility is placing the coils underneath the house during construction to gain heat loss from the house itself. Approximately 10 to 15 percent of the heat in a house is lost through conduction under the house.

Oklahoma is the center of earth-source pumps in the United States. Oklahoma State University sponsors the annual Heat Pump Technology Conference in Stillwater. The fifth conference was held in 1980 and the proceedings are available from the university. The School of Technology at the university has conducted detailed field demonstration experiments on horizontal earth coils.

Oklahoma University has designed and installed a vertical earth coil, like a closed well. The "well" is 240 feet deep and is made of 5-inch PVC pipe with a 1¼-inch pipe feeding the water from the heat pump down to the bottom of the closed well. As the water collects heat by conduction from the earth, it rises to the top where it is drawn off to supply the heat pump.

Installed cost for the drilling of the well and for the earth coil is $6.50 a foot; 240 feet cost $1,560. The two advantages of this system are that the earth-source temperature in a well can be as much as 20° F. warmer in winter than at a coil four feet from the surface. The other advantage is that there is less likelihood of leakage due to surface disturbance.

A solar-assisted system was tried with the intention of charging the well coil with summer heat for winter use, but during the night the temperature always returned to that of the surrounding earth.

James Parkin of Geosystems, Inc. (3623 North Park Drive, Stillwater, OK 74074) has installed nearly 100 earth-source heat pumps in the Stillwater area. His own 2,610-square-foot house has an 860-foot horizontal earth coil of four-inch diameter PVC pipe. The house is very well insulated. The cooling load is 31,788 Btuh at 100° F. and the heating load is 49,391 Btuh at 0° F. The heat pump

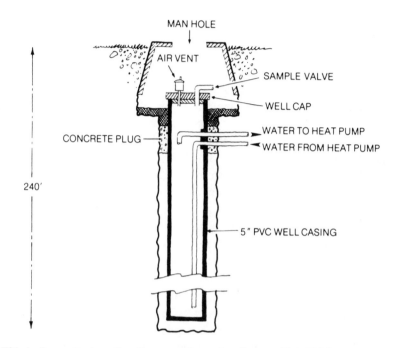

This is the vertical earth coil, or geothermal well, tested by Oklahoma University. Water is circulated in 240-foot well to pick up earth-source heat. As warm water rises to the top, it is carried to the heat pump.

is a three-ton, water-to-air unit. When the supply water temperature from the earth coil reaches 38° F., the heat pump is turned off automatically to prevent ice formation, and auxiliary resistance heat is turned on. A desuperheater provides Parkin with most of his domestic hot water. The COP on the system averages out at 2.5 and the SPF is 2.2.

The pipeline costs $2 a linear foot when installed with a septic lateral and $3 without. This amounts to between $1,680 and $2,520 for Parkin's 840 feet. The average water flow in the line is twelve gallons an hour. Some leaking did occur with his system. Leaking can be caused by heavy equipment passing over the pipe field, badly made connections, or even expansion and contraction. Sometimes water will surface at the leaking point but if not, it could be a very expensive repair job searching for the problem.

Tests of system

Phillips Company in Germany has tested an earth-to-air heat pump system. Its system has a COP of 3.5 and uses 400 feet of ground coil under the basement area of approximately 1,500 square feet. In winter, almost constant heat at 45° F. is extracted from the earth and pumped up to 120° F. for home heating.

The International Energy Agency is conducting experiments into earth-source heat pump systems in Germany, the Netherlands, and Sweden. All of these experimental systems are operative and are undergoing extensive testing.

The advantage of using an earth-source system is that the earth is a fairly constant source of heat. As a result, defrost cycles, which one must have with an air-source heat pump, are unnecessary. Also operating efficiencies increase because of the somewhat higher working temperature, and there is some slight additional gain in that less power is used to operate a small circulating pump than to operate the powerful fan required with a heat pump — particularly in summer. The disadvantages of using earth are the cost of laying coil, the possibility that the coil may leak, and the fact that one must use a water-to-air or water-to-water heat pump.

Earth-Source System

John Sumner is England's foremost exponent of the virtues of heat pumps, and he's probably had more experience with earth-source heat pumps than anybody else in the world. He has also written two books on the subject. From 1950 to 1954 he heated his house using air as the heat source; in 1955 he changed over to an earth-source heat pump which he has since used.

Sumner studied at Nottingham University under Professor Robinson, who had close associations with Lord Kelvin, the inventor of the heat pump. Sumner designed and built the first large heat pump in England for the Norwich Corporation electricity department.

I visited Sumner's house — the first and only house I've been in that was heated by a vegetable patch. The heat, taken out from three feet beneath the garden, is passed through a heat pump to underfloor heating inside the house. The 470 feet of copper ground coil were installed in 1955 and are spread over an area of 1,350 square feet of garden. The diameter of the pipe is one inch and provides 46.8 Btuh per linear foot of pipe at a ground temperature of 34.5° F.

It is remarkable to think that the cold, wet soil could possibly keep the house warm and dry, but there's no disputing the facts which Sumner has painstakingly recorded for over twenty-five years. The temperature of the antifreeze solution flowing in from the ground coil remains at a fairly constant 40° F. during winter. The temperature of the outgoing solution is 30° F. The two most interesting pipes in the heat pump are the outgoing one, which is coated in frost, and the pipe going to heat the house, which is 120° F. maximum. There can be no more graphic illustration of the heat pump than to put one hand on the freezing pipe and the other on the hot pipe.

John Sumner stands next to his Mark II heat pump. Sumner built the first large heat pump in England.

The 1,650-square-foot single-story house is thoroughly insulated. The underfloor heating system is serviced by 3/8-inch-diameter copper pipe buried at a depth of two inches from the top of the concrete floor. The pipes are spaced twelve inches apart throughout the house. The water is the house coil circulates at a maximum of 120° F. Surface temperature in the tiled floors is 85° F. and on the carpets it is 75° F. This provides a uniform house temperature of 68° F. The concrete floor in which the heating coil is embedded acts as a thermal store.

The heat load throughout the winter heating season averages at 12.5 kw. The SPF is maintained at three. I found the underfloor heating to be remarkably comfortable. The heat is more radiant than convected. As a consequence, there are fewer hot spots than one gets with air ducting or wet radiators. Also there are no panels or ducts in the house. During the six months' heating season he keeps his house at a constant 68° F., cooler than most centrally heated homes, but still quite adequate to provide comfortable background heating.

Sumner illustrates the psychology of heat as follows: Should a visitor comment on the low temperature, he will switch on a coal-effect fire, which gives an illusion of heat, and then after a while he casually asks if the guest is feeling any warmer. Invariably, he says, the answer is "yes." So heat is, to a small extent, relative. Needless to say, if one is still feeling cold, Sumner will turn on additional electric resistance heating.

Sumner designed and built his own original heat pump, the one that has successfully operated for twenty-five years. Last year he had a new one built, again to his own design, because he thinks his 1955 model may want to go into retirement soon.

To replace Sumner's system with copper pipes and a handmade heat pump would be costly. If PVC were used, one would need 940 linear feet (470 × 2), which at $3 a foot would cost $2,820. Installed, the heat pump should cost less than $2,500. Geosystems, Inc. is a source for a ground coupling type of unit. The cost of the indoor heating system would depend on whether standard air duct heating or underfloor heating was chosen. The cost of the latter would depend on whether copper pipe or plastic garden hose was used.

So there it is — a house that has been heated with the earth since 1955 and continues to function very well.

SUMMARY

Location: Norwich, England
Date of completion: 1955
System: Earth-to-underfloor heating
                                  using water-to-water heat
                                  pump
Heat pump: Individual design and handmade
House size: 1,650 square feet
SPF: 3
Cost: Replacement cost $5,500 exclud-
                                  ing underfloor heating system

PART III

APPENDIX

"Even when the air is below 32 F., it contains vast reserves of heat that can be used by a heat pump."

APPENDIX A

Thermodynamics Of Heat Pumps

Figure 1 shows the elements of a heat pump using a liquid refrigerant in a vapor compression cycle. At point D the refrigerant is a liquid whose temperature is lower than that of the heat source (air, water, sun, or earth). Heat, therefore, transfers from the source to the refrigerant. Because of this transference as well as a reduction in pressure due to the suction of the compressor, the liquid changes state. It becomes a vapor at point A (i.e., it boils) and in so doing absorbs the heat necessary to supply the latent heat of vaporization. This occurs at constant temperature, T_1, and an amount of heat, Q_1, is absorbed from the heat source.

The vapor is now compressed in the compressor and its temperature raised to T_2 with the absorption of the heat equivalent of the mechanical work input Q_w. From point B it then passes through the condenser and gives up its latent heat to the heat exchanger medium (air, water) and in so doing condenses to a liquid at high pressure. From C it passes through the expansion valve, causing the temperature and pressure both to drop.

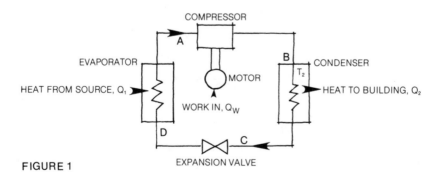

FIGURE 1

The coefficient of performance is defined by:

$$COP = \frac{\text{heat delivered}}{\text{work input}} = \frac{Q_2}{Q_w}$$

The most commonly used diagram for showing the operation of heat engines (a heat pump is essentially a heat engine "in reverse") is the temperature versus entropy graph. ("Entropy" is defined as enthalpy, or quantity of heat, divided by absolute temperature, i.e., $S = Q/T$.)

The standard reference cycle for heat engines is the Carnot cycle shown in its reverse form in Figure 2. From D to A, a quantity of heat, Q_1, is absorbed by the medium isothermally, (i.e., at constant temperature T_1), where $Q_1 = (S_2 - S_1)T_1$.

From A to B the medium is compressed, its temperature raised to T_2, and a quantity of heat, $Q_w = S_2(T_2 - T_1)$, added due to the mechanical work input to the compressor. From B to C a quantity of heat, Q_2, is given up, isothermally, where $Q_2 = (S_2 - S_1) T_2$. Finally, from C to D, the medium expands either through an expansion valve or by working a turbine which could provide some of the work required by the compressor. In either case, energy is given out equivalent to a quantity of heat, $Q_e = S_1 (T_2 - T_1)$.

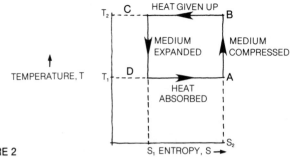

FIGURE 2

The theoretical maximum COP for the cycle is found when the energy ouput from cycle section C to D is used as input to the work input during section A to B, i.e., equal to $Q_w - Q_e$. The COP is then found as follows:

$$COP = \frac{\text{heat delivered}}{\text{work input}}$$

$$= \frac{Q_2}{Q_w - Q_e}$$

$$= \frac{(S_2 - S_1)T_2}{S_2(T_2 - T_1) - S_1(T_2 - T_1)}$$

$$= \frac{T_2}{T_2 - T_1}$$

In practice, heat pumps do not operate according to the theoretical reverse Carnot cycle for a number of reasons, some of the more important of which are:

1. The heat emitted during the expansion, Q_e (e.g., through an expansion valve as described above for a practical vapor compression cycle) is not used to offset the work input, Q_w, but is lost;
2. The heat absorption and emission are not truly isothermal because the heat exchange medium changes temperature as heat is gained or lost;
3. The compression and expansion are not truly adiabatic in practice ("adiabatic" means no heat is lost to or gained from cylinder walls and pistons or by friction).

The loss of heat to compresser and heat exchangers alone can account for a 50 percent fall-off in achievable COP so that the achievable COP is no more than $0.5 \dfrac{T_2}{T_2 - T_1}$. The loss of heat from the expansion valve used in the vapor compression cycle will cause a further loss in the order of 15 percent so that the theoretically achievable COP can become

$$0.5 \times 0.85 \frac{T_2}{T_2 - T_1} = 0.42 \frac{T_2}{T_2 - T_1}.$$

In practice, a number of different cycles are used by designers but they will not be gone into here. Suffice to say that the theoretical maximum Carnot cycle COP ($\frac{T_2}{T_2 - T_1}$) can be used as a standard, but even the best practical cycle could not be expected to achieve an ideal COP of more than half this value.

APPENDIX B

Heat Pipes

The heat pipe (a term coined in 1963) is another example of the heat pump principle in operation, in that it follows a similar cycle of evaporation and condensation. The main difference is that no compressor is used to "pump up" the heat to a useful temperature.

The heat pipe is a closed tube. At one end where the heat is, the liquid (methanol or Freon) boils and evaporates, causing an increase in pressure which drives gas up the hollow center. On reaching the cooler end, the gas condenses to a liquid and so gives up its heat. The liquid is then drawn down through a porous wick to replace that which was evaporated. As long as there is a temperature difference between the two ends, the cycle will continue.

As a straightforward heat exchanger, the heat pipe is excellent for the following reasons:

1. It operates without the need for any mechanical energy (a pump is required with a solar collector to shift the hot water from the panel to the tank).
2. The heat pipe is a remarkably good conductor of heat.
3. Despite the fact that it absorbs heat at one end and gives it out at the other, there is almost no temperature difference inside the heat pipe.

As a result of these unusual qualities, the heat pipe has recently become of use in such diverse areas as spacecraft and thermogrates (heat-conducting fireplaces used for hot water and central heating).

APPENDIX C

Heat Pumps and Tax Credits

There is no doubt but that the use of heat pumps may reduce the amount of oil, gas, coal, or electricity you use to heat or cool your home.

There is, too, no doubt but that heat pumps do *not* qualify for Federal energy credits for income tax purposes.

The Internal Revenue Regulation 1.44C-2(e) states, "Thus, *heat pumps* or oil or gas furnaces, used in connection with renewable energy source property, are not eligible for the credit." The regulation defines renewable energy source property as solar, wind or geothermal energy property.

The one installation that could qualify for the energy credit is a solar system with a heat pump used to boost the heat of the solar system to a more useful level. In this case, the cost of the solar system, but not the heat pump, would qualify.

Federal laws do change, however, and so do the interpretations of the federal regulations. If you believe your heat pump installation should qualify, since it conserves fossil fuel, discuss your installation with the IRS representative in your area.

APPENDIX D

Heat Storage

The advantage of using water as a heat storage medium is its high specific heat, i.e. the amount of heat required to raise the temperature of a substance by a specified amount. The British Thermal Unit (Btu) is defined as the amount of heat required to raise the temperature of one pound of water by 1° F. The specific heat of water in this system of units is, therefore, 1 Btu/lb.

Water has the highest specific heat of any known substance, liquid or solid. The following is a comparison of the specific heats of a few common substances:

Water	1 Btu/lb.
Oil	0.4
Air (normal temp.)	0.24
Aluminum	0.214
Steel	0.117
Lead	0.030
Concrete	0.19
Glass	0.19
Wood (average)	0.54

One U.S. gallon of water weighs 8.35 lbs. so that one gallon would require 8.35 Btu to raise its temperature by one degree. The quantities are related by the single formula:

Heat required to raise temperature of water (Btu)
= weight of water (lbs.) × temperature rise (° F)
= volume (U.S. gallons) × 8.35 × temperature rise (° F).

For example, suppose we have a 1,000-gallon water tank and we want to raise its temperature from 70° F. to 120° F. The heat required to do this will be 1,000 × 8.35 × (120 ¹ 70) = 47,500 Btu.

If the capacity in gallons of a water tank is not known, it can be readily obtained from the dimensions since one cubic foot is equivalent to 7.48 U.S. gallons (6.22 British or Imperial gallons). One cubic foot of water weighs 62.4 pounds. A tank 4, × 3, × 5, for example, will hold 60 cubic feet or 448.8 gallons, and if the liquid is water, this will weigh 3,747 pounds.

Since heat is required to raise the temperature of the water, this heat is effectively stored as long as the temperature remains at the higher value. As the heat is released to, for example, the surrounding lower temperature air, the temperature of the water drops. Since the rate at which heat can flow from one object to another depends on the temperature difference between them, the higher the temperature of the medium storing the heat, the higher will be the rate of heat exchange.

Storage tanks

Steel tanks are expensive to use as water storage tanks. They should be located above ground because they are subject to severe corrosion below ground unless protected with special inhibitors. The tanks must be kept full, but if they are not they should be equipped with an expansion tank vented to the outside to prevent vapor space corrosion. Interior corrosion problems with hot water storage necessitate special coatings with materials like rubber or plastic liners, plastic paint-like products, and galvanizing.

Concrete tanks for thermal water storage are available in different types: reinforced concrete blocks, precast utility vaults, and precast storm drain pipes. Pre-stressed, precast tanks are an economical way to store water.

Avoid integration of concrete tanks with the house structure because of thermal expansion. As with all storage systems, be sure the floor can support the weight. Storm drain pipe is ideal because it is made on a regional basis in sizes up to eight feet in diameter. It is overdesigned to take road traffic loads. The joints should be sealed with synthetic rubber.

All concrete tanks should be internally coated or lined to prevent buildup of alkaline solids in the water from the concrete. The most suitable coating able to take high storage temperatures is phenolic resin applied over a sand-blasted or acid-etched surface.

Wood tanks are also suitable but they must be kept totally dry or totally wet for durability — the rot occurs in in-between stages.

Fiberglass tanks are available and have excellent durability. When buying a fiberglass tank, be sure that it is designed to withstand high temperatures. For example, if the thermal store is designed never to exceed 150° F., tanks used for oil storage are suitable, but if the temperature is expected to exceed that, a special tank is required. Fiberglass tanks are available for temperatures up to 240° F.

Finally, various types of pillow storage systems are available using insulated plastic or rubber sacks that can be installed in a crawl space.

Rock storage

The advantage of rock storage over water is that it does not leak, corrode, or rot. The disadvantage is that the rock store needs to be about 2.5 to 3 times the size of the water store to hold as much heat.

The rocks should be one to two inches in diameter. They should be river rock, washed to remove contaminants and dust and kept dry in the bin so that no mold or mildew occurs. Stones should be nonporous and should not give rise to fine dust over time. Smooth quartz river gravel is the best choice. A plenum should be left above and below the bin to allow circulation of air. One system, installed by Westinghouse, uses concrete blocks at the bottom of the store. The blocks are spaced two inches apart to allow air to circulate, and several layers of half-inch hardware cloth are placed over the blocks to prevent the stones from falling down and clogging the space between the blocks.

The installation uses a five-foot diameter by eight-foot high steel culvert containing 15,000 pounds of stones which can store about 150,000 Btu. It will provide overnight heating in the cold winter period and a few days' heating during the milder fall and spring.

Phase-change materials

Phase-change materials, such as paraffin and Glaubers' salts, absorb heat by changing into a liquid and yield it by changing back into a solid. The advantage is compactness. Only a half or quarter of the space needed for water storage is required. The disadvantages are higher costs and that long-term reliability is not known.

All thermal stores should be insulated to a value of between R-19 and R-40. If buried in ground the insulation should be polyurethane, styrofoam, or fiberglass with a vapor barrier.

APPENDIX E

Refrigerants

The earliest vapor compression cycle machines used air as the heat transfer medium. Refrigerants were developed to replace air. They give a far better heat transfer efficiency. There are any number of refrigerants to choose from but the most popular in use today for domestic heat pumps are R-12 and R-22, also known as Freon 12 and Freon 22 (Freon is a Du Pont Trademark). The advantage refrigerants have over air is their low boiling temperature, for example, R-12 boils at $-21.7°$ F. The low boiling temperature increases the heat absorption ability of the refrigerant as it changes state from a heavy vapor to a gas. The most common refrigerants used in the vapor compression cycle are listed below.

REFRIGERANT NO.	NAME AND CHEMICAL FORMULA
R-11	Trichloromonofluoromethane CCl_2F
R-12	Dichlorodifluoromethane CCl_2F_2
R-22	Monochlorodifluoromethane $CHClF_2$
R-500	Azeotropic mixture of 73.8% of (R-12) and 26.2% of (R-152a)
R-502	Azeotropic mixture of 48.8% of (R-22) and 51.2% of (R-115)
R-503	Azeotropic mixture of 40.1% of (R-23) and 59.9% of (R-13)
R-504	Azeotropic mixture of 48.2% of (R-32) and 51.8% of (R-115)
R-717	Ammonia NH}

The Market Potential

The heat pump industry has a strong and exciting future. As our non-renewable fuels become costlier and availability less certain, increasing attention will be paid to the sources of "free" heat that surround us: heat in the air, water, sun, and earth. Heat pumps, which use this free heat, have already produced efficiencies which far exceed today's conventional fuels. But this is only the beginning.

In domestic heat pumps, we will see the specialization and adaptation of the product to suit different climates. Efficiency ratings could be greatly increased if, for example, there were one type of heat pump for Vermont with its long, cold winters, and another type for Florida with its mild winters.

The market for retrofitting existing homes with heat pumps will continue to expand. And there is the untapped market for retrofitting heat pumps and air conditioners with desuperheating kits. Hotels, restaurants, and other commercial establishments, as well as homes could make good use of this now-wasted heat.

No one has thought to adapt heat pumps to the specialized needs of laundromats and laundries for hot air, hot water, and cold rinsing water. The food processing industry has a voracious appetite for heating and cooling both air and water, and in this lies not only great potential for heat transfer but also for heat pumps.

And there is potential application of heat pumps (and thermal compressors) in agriculture.

Maybe the greatest potential is in coupling heat pumps and solar energy. This can be done successfully by using passive solar principles to increase efficiency on both the heating and cooling cycle, and by converting the outdoor heat exchange coil to a refrigerant-charged solar collector. These measures could increase efficiencies by 50 to 200 percent.

New construction, both domestic and commercial, is bound to move towards passive solar gain. A small heat pump is perfect for balancing the heating and cooling load in passive solar homes and for providing hot water.

The opportunities are endless; the next twenty-five years are certain to be active, exciting years in the heat pump industry.

Manufacturers' Index

Addison Products Company
P.O. Box 63
Addison, MI 49220

Manufactures domestic and commercial heat pumps. Has ten different models of WeatherKing air-source heat pumps in either packaged or split systems. The models have COP's ranging from 2.0 to 2.9 (47° F.) and 1.5–2.1 (17° F.). Also has six models of WeatherKing Add-A-Pump. These are air-source, split systems with a COP of 2.6–2.7 (47° F.) and 1.7–2.0 (17° F.).

Air Conditioning
P.O. Box 6225
Greensboro, NC 27405

Water-source heat pumps.

Airtemp Corporation
Woodbridge Ave.
Edison, NJ 08117

Has air- and water-source heat pumps. The five air-source, split-system models have a COP of 2.3–2.7 (47° F.) and 1.6–1.8 (17° F.). The outdoor unit of these pumps is available as part of a solar-assisted system.

185

Amana Refrigeration, Inc.
Amana, IA 52203

Firm has packaged and split-system, air-source heat pumps with COP's of 2.4–2.9 (47° F.) and 1.5–1.8 (17° F.).

American Air Filter Company Inc.
13353 Alondra Blvd.
Santa Fe Springs, CA 90670

Manufactures residential and industrial heat pumps using water, air, or sun as the heat source. Fourteen models of packaged, water-source pumps are marketed under the trade name of EnerCon and have COP's ranging from 2.5–3.1.

American Solar King Corporation
6801 New McGregor Highway
Waco, TX 76710

Markets five models of the Solar King, a packaged water-source heat pump, with COP's ranging from 2.7 to 3.1.

Bard Manufacturing Company
P.O. Box 607
Bryan, OH 43506

Offers a variety of air-source heat pumps in both packaged and split systems. The COP's range from 2.0–3.1 (47° F.) and 1.3–2.2 (17° F.).

Barkow Manufacturing Company Incorporated
2230 S. 43rd St.
Milwaukee, WI 53219

Manufactures air- and water-source heat pumps.

B.D.P. is an amalgamation of three corporations:

Bryant Air Conditioning
7310 W. Morris St.
Indianapolis, IN 46231

Day & Night Air Conditioning
855 Anaheim-Puente Rd.
La Puente, CA 91749

Payne Air Conditioning
855 Anaheim-Puente Rd.
La Puente, CA 91749

The product range for the three corporations is the same. They handle split and packaged, air-source heat pumps with COP's of 2.3–3.0 (47° F.) and 1.5–2.0 (17° F.). B.D.P. also offers a solar heat pump system.

Bryant Air Conditioning, *see* **B.D.P.**

Calmac Manufacturing Company
150 Brunt St.
Englewood, NJ 07631

Makes water-source heat pumps.

Carrier Air-Conditioning Company
Carrier Pkwy.
Syracuse, NY 13201

Manufactures six models of the Weathermaker, a water-source, packaged heat pump, with COP's of 2.6–3.1. The air-source Weathermaster III's have COP's of 2.9–3.0 (47° F.) and 2.0–2.1 (17° F.). The air-source Weathermakers, available packaged (four models) or split system (four models), have COP's of 2.7–2.9 (47° F.) and 1.7–2.1 (17° F.). They come with a solar option. Carrier also manufactures the "Hot Shot," a domestic hot water preheater which will supply as much domestic hot water from the heat pump as an "average" solar panel system.

Century By Heat Controller
1900 Wellworth Ave.
Jackson, MI 49203

Manufactures air-source heat pumps.

Coleman Company, Inc.
250 North St., Francis St.
Wichita, KS 67201

The Pres. II Heat Pump D.E.S., made by the Coleman Company, is an air-source, split system. The unit is available in five models with COP's of 2.9–3.1 (47° F.) and 2.3–2.4 (17° F.).

Comfort Enterprises Company
Div. of Herrmidifier Company, Inc.
P.O. Box 323
Leola, PA 17540

Manufactures air-source heat pumps.

Command-Aire Corporation
3221 Speight Ave.
Box 7916
Waco, TX 76710

Manufactures a variety of air- and water-source, as well as solar-assisted, heat pumps. The COP's range from 3.0–3.6 for the packaged, water-source units.

Crane Supply Company
300 Park Ave.
New York, NY 10022

Manufactures air-source heat pumps.

Day & Night Air Conditioning, *see* **B.D.P.**

Dunham-Bush, Inc.
Harrisonburg Division
101 Burgess Rd.
Harrisonburg, VA 22801

The Dunham Bush air-source heat pumps are split systems with COP's of 2.2–2.5 (47° F.) and 1.7–1.8 (17° F.). The company also manufactures large industrial and commercial units.

Dunham-Bush, Inc.
175 South St.
West Hartford, CT 06110

Carries water-source heat pumps.

Duo-Therm
509 S. Poplar St.
La Grange, IN 46761

Manufactures air-source heat pumps.

Eubank Manufacturing Enterprises, Inc.
FM 2011 South, I-20
Longview, TX 75601

Offers air-source heat pumps.

Fasco Industries Inc.
810 Gillespie St.
Fayetteville, NC 28302

Manufactures air-source heat pumps.

Fedders Corporation
Woodbridge Ave.
Edison, NJ 08817

Has many models of water- and air-source heat pumps. The Adaptomatics are their air-source packaged units and have COP's of 2.3–2.7 (47° F.) and 1.6–1.8 (17° F.); the Thermizers are air-source, split-system heat pumps with COP's of 2.2–2.8 (47° F.) and 1.6–2.0 (17° F.). Fedders manufactures solar-assisted heat pump systems using these units.

Florida Heat Pump Corporation
Leigh Products Inc.
610 S.W. 12th Ave.
Pompano Beach, FL 33060

Markets many models of the Energy Miser, a water-source heat pump with COP's of 2.7–3.5. All units are available as packaged or split-system heat pumps.

Friedrich Air Conditioning & Refrigeration Company
4200 N. Pan Am IH 35 Expressway
San Antonio, TX 78295

The Friedrich-Climate Master Series offers a wide variety of packaged, water-source heat pumps with COP's of 2.6–3.7. The Friedrich air-source models have COP's of 2.1–2.9 (47° F.) and 1.3–2.0 (17° F.). These are available as packaged or split systems. Friedrich also manufactures commerical and industrial heat pump systems, as well as the "Hot Water Generator" desuperheater.

General Electric Company
Troup High
Tyler, TX 75701

Weathertron by G.E. is an air-source, packaged or split-system heat pump. It is available in many models with COP's of 2.2–2.9 (47° F.) and 1.4–2.1 (17° F.).

Gervais Equipment
9295 Fargo Rd.
Stafford, NY 14143

Makes water-source heat pumps.

Goettl Air Conditioning, Inc.
2005 E. Indian School Rd.
Phoenix, AZ 85016

Has air-source, packaged or split-system heat pumps with COP's of 2.3–2.7 (47° F.) and 1.5–1.8 (17° F.).

Heat Controller, Inc.
Losey at Wellworth
Jackson, MI 49203

Comfort-Aire, an air-source heat pump made by Heat Controller, is available either packaged or split system. Within the ten models, the range in COP's is 2.0–2.9 (47° F.) and 1.5–2.1 (17° F.).

Heat Exchangers, Inc.
8100 N. Monticello Ave.
Skokie, IL 60076

Heat Exchangers, Inc. markets eleven models of Koldwave, a packaged, water-source heat pump with COP's between 2.6–3.1.

Heil-Quaker Corporation
647 Thompson Lane
Nashville, TN 37204

The Heil air-source heat pumps have COP's of 2.6–2.9 (47° F.) and 1.7–2.1 (17° F.). They are available either packaged or split system.

Henry Furnace Company, *see* **Luxaire, Inc.**
Medina, OH 44256

Herrmidifier Company, Inc.
1810 Hempstead Rd.
P.O. Box 1747
Lancaster, PA 17604

Manufactures air-source heat pumps.

Home Division, L.S.I.
900 Brooks Ave.
Holland, MI 49423

Manufactures air-source heat pumps.

International Environmental Corporation
P.O. Box 25608
Oklahoma City, OK 73125

Paramount Efficiency by International Environmental Corporation is available in six models. They are water-source, packaged systems with COP's of 3.3–3.5.

Intertherm Inc.
3800 Park Ave.
St. Louis, MO 63110

Intertherm manufactures air-source heat pumps.

Janitrol, a Tappan Division
400 Dublin Ave.
Columbus, OH 43216

The Janitrol packaged and split, air-source heat pumps have COP's of 2.3–2.8 (47° F.) and 1.3–2.1 (17° F.). One of the split-system models is for heat only and has COP's of 3.2 (47° F.) and 1.9 (17° F.).

Lau Industries
2027 Home Ave.
Dayton, OH 45407

Makes air-source heat pumps.

Lennox Industries, Inc.
200 S. Twelfth Ave.
Marshalltown, IA 50158

Heat pumps by Lennox Industries are sold under two trade names. Both use air as the heat source. Solarmate pumps are available as packaged or split systems and have COP's of 2.1–2.8 (47° F.) and 1.6–1.9 (17° F.). (The air-source, packaged units have the solar option.) Fuelmaster is available only as a split system. These models have COP's of 2.5–2.9 (47° F.) and 1.9–2.1 (17° F.).

Lowe's Company, Inc.
Box 1111
N. Wilkesboro, NC 28659

The air-source heat pumps manufactured by Lowe's Company are available either as packaged or split systems and have COP's of 2.2–2.9 (47° F.) and 1.3–1.9 (17° F.).

Luxaire, Inc.
West of Fibert St.
Elyria, OH 44035

There are many models of Luxaire air-source heat pumps. These packaged and split-system units have COP's of 2.0–3.0 (47° F.) and 1.2–1.9 (17° F.).

Magic Chef, Inc.
851 W. Third Ave.
Columbus, OH 43212

Magic Chef has air-source, packaged and split-system heat pumps with COP's of 2.2–2.8 (47° F.) and 1.2–1.9 (17° F.).

Mammoth
Div. of Lear Siegler, Inc.
941 E. 7th St.
Holland, MI 49423

Mammoth has many models of their water-source, packaged heat pump, with COP's ranging from 2.7–3.2. They also manufacture the Hydrobank heat pump system for large scale use.

Marvair Company
P.O. Box 400
Cordele, GA 31015

The Marvair heat pumps are air-source, packaged or split-system units with COP's of 2.3–2.6 (47° F.) and 1.7–1.9 (17° F.). Marvair also makes a desuperheater.

McDonald Manufacturing Company
P.O. Box 508
Dubuque, IA 52003

Manufactures air-source heat pumps.

McGraw-Edison Company
Air Comfort Divison
704 N. Clark St.
Albion, MI 49224

Manufactures air-source heat pumps.

McQuay Division
McQuay-Perfex Inc.
13600 Industrial Park Blvd.
Minneapolis, MN 55440

Manufactures both air- and water-source heat pumps.

Melnor Industries
1 Carol Place
Moonachie, NJ 07074

Manufactures air-source heat pumps.

Melnor Manufacturing, Ltd.
80 Morton Ave.
E. Bransford, Ont., Canada

Manufactures air-source heat pumps.

Montgomery Ward & Company, Inc.
P.O. Box 8339
Chicago, IL 60680

Supplies air-source heat pumps in split-system or packaged units. They have COP's of 2.6–2.7 (47° F.) and 1.8–2.0 (17° F.).

Mueller Climatrol Corporation
Woodbridge Ave.
Edison, NJ 08817

Makes water-source heat pumps.

Northrup Inc.
302 Nichols Dr.
Hutchins, TX 75141

Makes the Energy Exchange Systems which are water-source, packaged heat pumps with COP's of 2.7–3.2.

O.E.M. Products, Inc.
2701 Adamo Dr.
Tampa, FL 33605

Manufactures air- and water-source heat pumps as well as solar-assisted pumps.

Patco, Inc.
6955 Central Hgwy.
Pennsauken, NJ 08109

The Patco air-source heat pumps have COP's of 2.2–2.8 (47° F.) and 1.2–2.0 (17° F.). They are available as packaged or split-system units.

Payne Air Conditioning, *see* **B.D.P.**

Phoenix Air Conditioning, Inc.
651 Vernon Way
El Cajon, CA 92020

Manufactures five models of their water-source heat pump in packaged or split systems. Desuperheaters are an integral part of the units.

Rheem Manufacturing Company
Air Conditioning Division
5600 Old Greenwood Rd.
Fort Smith, AK 72901

There are many models of the air-source, packaged or split-system heat pumps made by Rheem Manufacturing. The COP range of these units is 2.3–3.2 (47° F.) and 1.5–2.3 (17° F.).

Sears, Roebuck & Company
925 S. Homan Ave.
Chicago, IL 60607

The Sears air-source heat pumps have COP's of 2.3–2.9 (47° F.) and 1.6–2.1 (17° F.) and are available as packaged or split systems.

The Singer Company
Climate Control Division
1300 Federal Blvd.
Carteret, NJ 07008
        or
1300 Hampton Ave.
St. Louis, MO 63139

The many models of the Electro Hydronic, manufactured by The Singer Company, are water-source, packaged units with COP's of 2.7–3.0. Singer also manufactures AFCO Comfortmakers. They are air-source heat pumps (packaged or split system) with COP's of 2.3–2.8 (47° F.) and 1.6–1.9 (17° F.).

Skuttle Manufacturing Company
Route 1
Marietta, OH 45750

Manufactures air-source heat pumps.

Solar Energy Research Corporation
1224 Sherman Dr.
Longmount, CO 80501

Makes water-source and solar-assisted heat pumps.

Solar Oriented Environmental Systems, Inc.
10639 S.W. 185th Ter.
Miami, FL 33157

The SOESI Power Savers are water-source, packaged heat pumps with COP's of 2.7–3.2. All units include an integral desuperheater. SOESI also manufactures industrial and commerical heat pumps.

Solartherm Manufacturing Corporation
768 Vella Rd.
Palm Springs, CA 92262

Manufactures air- and water-source heat pumps and solar-assisted pumps.

Southwest Manufacturing
Division of McNeil Corporation
30 No. Elliott
Aurora, MO 65605

Has two models of Heatwave, an air-source, split-system heat pump with COP's of 2.6–2.7 (47° F.) and 2.0 (17° F.)

Spartan Electric Company
P.O. Box 150
Fayetteville, NC 28302

Spartan heat pumps are air-source, packaged or split-system units with COP's of 2.4–2.7 (47° F.) and 1.4–1.8 (17° F.).

Spectrum Solar Systems Corporation
11615 Saylor Rd.
Pickerington, OH 43147

Manufactures water-source heat pumps.

Square D Company
P.O. Box 766
Mesquite, TX 75149

Has two models of an air-source, split-system heat pump. The units have COP's of 2.3 (47° F.) and 1.8–1.9 (17° F.).

Tappan Company
Air Conditioning Division
206 Woodford Ave.
Elyria, OH 44035

Manufactures packaged and split-system, air-source heat pumps with COP's of 2.3–2.8 (47° F.) and 1.3–2.1 (17° F.).

Tempmaster International
1775 Central Florida Pkwy.
Orlando, FL 32809

Manufactures water-source heat pumps.

Thermal Energy Transfer Corporation
378 W. Olentangy St.
P.O. Box 397
Powell, OH 43065

Makes water-source heat pumps.

The Trane Company
3600 Pammel Creek Rd.
La Crosse, WI 54601

Produces packaged and split-system, air-source heat pumps with COP's of 2.2–3.0 (47° F.) and 1.5–1.8 (17° F.). Also markets an add-on unit.

Typhoon Air Conditioning Company
1135 Ivanhoe Rd.
Cleveland, OH 44110

Makes air-source heat pumps.

Vanguard Energy Systems
9133 Chesapeake Dr.
San Diego, CA 92123

The water-source, packaged heat pumps produced by Vanguard Energy Systems have COP's of 2.91–3.62 at 70° F. Vanguard also manufactures a desuperheater kit, as well as an add-on series of water-source heat pumps and solar heat pump systems.

Vilter Manufacturing Corporation
2217 S. First St.
Milwaukee, WI 53207

Manufactures water-source heat pumps.

Walton Laboratories, *see* **Melnor Industries and Melnor Manufacturing Ltd.**

WeatherKing, Inc., *see* **Addison Products Company**

Wescorp, Inc.
15 Stevens St.
Andover, MA 01810

Makes water-source heat pumps.

Westinghouse Electric Corporation
P.O. Box 1283
Norman, OK 73069

Manufactures four models of air-source heat pumps. There are three sizes of the WhispAir, a packaged unit. The Westinghouse is available either in the packaged or split system; the Hi/Re/Li and the Thriftaire are split systems. The COP's of these different models range from 2.3–2.9 (47° F.) and 1.4–2.1 (17° F.) Westinghouse's line of heat pumps can be adapted to their "Native Sun" solar heat pump system. Westinghouse also manufactures the Templifier industrial heat pump.

Whirlpool Heating & Cooling Products, *see* **Heil-Quaker Corporation.**

Wilcox Manufacturing Corporation
13375 U.S. 19 North at 62 St.
P.O. Box 455
Pinellas Park, FL 33565

Manufactures water-source heat pumps.

The Williamson Company
3500 Madison Rd.
Cincinnati, OH 45209

Carries many different models of an air-source, split-system heat pump; also has three models of a packaged, air-source unit. The COP range is 1.8–2.8 (47° F.) and 1.2–2.0 (17° F.). Williamson also manufactures an add-on heat pump.

York Division of Borg-Warner Corporation
P.O. Box 1592
York, PA 17405

The York is packaged unit; the Champion is a split system. Both models are air-source heat pumps with COP's of 2.4–3.1 (47° F.) and 1.3–2.1 (17° F.). Add-on systems are available for the Champion.

John Zink Company
P.O. Box 7388
Tulsa, OK 74105

Manufactures air-source heat pumps.

Index